Geology of the continental margins

Geology of the continental margins

G. Boillot
Université Pierre et Marie Curie, Paris

Translated by A. Scarth
University of Dundee

Longman
London and New York

Longman Group Limited
Longman House
Burnt Mill, Harlow, Essex, UK

*Published in the United States of America
by Longman Inc., New York*

English translation
© Longman Group Limited 1981

First published as *Géologie des marges continentales* by Masson S. A., Paris in 1978
© Masson, Publisher, Paris, 1978
English edition first published in 1981

British Library Cataloguing in Publication Data

Boillot, G
  Geology of the continental margins.
  1. Continental margins
  2. Submarine geology
  I. Title   II. Scarth, A
  551.4'608   GC84   80-40317

  ISBN 0-582-30036-3

Printed in Great Britain by
William Clowes (Beccles) Ltd,
Beccles and London

Foreword

It is not very prudent to write a book in 1978 about the continental margins. The amount of research work in this area is such that every year brings out much new information and anyone who attempts a clarifying summary runs the risk of presenting his readers with a book that will be out of date very quickly. I took this risk because I sensed that the students felt an urgent need for a synthesis – however imperfect it may be – on a subject about which practically nothing was known fifteen years ago. It seemed to me also that many of my geologist colleagues in the universities or in industry, whose specialisms are far removed from geological oceanography, are also keen to know more about these continental margins where structural geologists see the cradle of the fold mountains and where the oil companies hope to find our hydrocarbon supplies for the next decade.

This book is therefore aimed at the reader who would like a broad presentation. This explains its brevity. It is an *introduction* to the geology of the continental margins, where regional descriptions are avoided in favour of the analysis of geodynamic phenomena and the description of explanatory models. I hope in this way to help students gain access to more specialised studies, whilst at the same time facilitating the dialogue between marine geologists and continental geologists.

I have benefited in the course of my work from the assistance of many colleagues and friends: J. Aubouin, G. Bellaiche, J. Dubois, R. Dubois, M. Gennesseaux, J. Hernandez, J. Lameyre, Y. Lancelot, X. Le Pichon, C. Lepvrier, J. Malod, J. Mascle, who is the initiator of this book, A. Mauffret, D. Mougenot, J.-P. Rehault, C. Robin, and, especially, J. R. Vanney, who have read and criticised my first manuscript or some of its chapters. I thank them warmly, and also G. Enard, J. Gosselin and L. Hoinard who arranged the text and its illustrations.

Contents

List of figures

(Full details of works referenced in the captions of these figures can be found on pp. 109–11.)

Acknowledgement

We are grateful to the following for permission to reproduce copyright material:

Princeton University Press for Fig. 1 p. 75 'Magnetic Anomalies Associated with Mid-Ocean Ridges' by F. J. Vine in *The History of the Earth's Crust* edited by Robert A. Phinney copyright © 1968 by Princeton University Press.

Chapter 1
General introduction: ocean and continent.
Plate tectonics

This is an introductory chapter. It provides a brief summary of the morphology of the oceans and the continental margins (§1). Then the structure of the continental crust is compared and contrasted with that of the oceanic crust and the idea of the intermediate crust on the stable margins is thereby introduced (§2). Finally, the lithosphere is defined in relation to the asthenosphere (§3) and the theory of plate tectonics is summarised in the last paragraph (§4)

The coast, for a geologist, is not the true boundary between continent and ocean. The rocks and structures on land seen by the geologist extend under the sea and the coastline rarely coincides with a geological boundary. The present level of the oceans is partly the result of the melting of the Quaternary glaciers and, indeed, during the cold periods of the Pleistocene, the sea retreated far across the continental shelf, sometimes as far as the shelf-break that now occurs at a depth of 130–200 m. This instability of the coastline has been a permanent feature throughout geological time, and the real boundary between the continents and oceans must be looked for far out to sea.

A brief study of the frequency curve of heights and depths on the surface of the globe shows a bimodal distribution (Fig. 1.1). The most frequent height is 300 m, the most

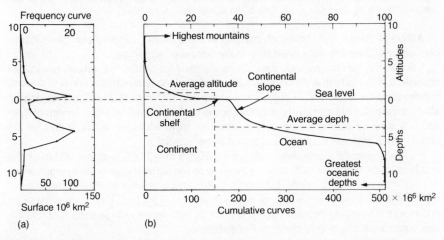

Fig. 1.1 Hypsometric curve of the earth's crust. (a) frequency curve; (b) cumulative curve; vertical scale in km. (After Wyllie, 1971.)

frequent depth is 4 800 m, whilst ocean floors are least frequent between 2 000 and 3 000 m. The surface of our planet thus very clearly reflects the basic duality of the crust: the land masses and the zones of the deep oceans belong to markedly different domains. The continental margins form a narrow transition zone whose characteristics and geological history are necessarily determined by the geodynamic factors belonging both to the continent and to the ocean.

1 The major morphological provinces of the sea-floor (Fig. 1.2)

(a) Active margins and stable margins

Two major categories of *continental margins* may be distinguished. Some margins are called *'active'* because they are characterised both by strong seismic activity and vigorous vulcanicity. They are chiefly found around the Pacific, where they form an almost continuous belt, and are often bordered by deep trenches. Other margins are called *'stable'* (or 'passive') because they betray no hint of any tectonic activity. This is the case on most of the Atlantic margins.

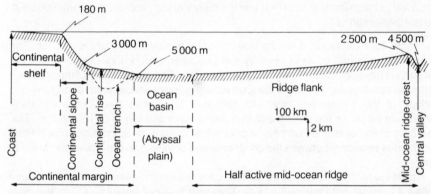

Fig. 1.2 The major morphological provinces of the sea-floor. Synthetic and schematic cross-section at right angles to the axis of an active mid-ocean ridge.

Although these two types of margin are the result of different geodynamic phenomena, they are nevertheless usually described together:
  • Bordering the land masses the *continental shelf* (or platform) is relatively wide but shallow. It is 70–80 km wide on average and occasionally reaches a width of several hundred kilometres, whilst its depth varies from 0 to 130 m or 180 m. Every geological study shows that it is merely the slightly submerged continuation of the continent from which it may be distinguished only by the thin layer of water above it and by the sedimentary features caused by recent marine activity.
  • The *continental slope* descends from −200 m to −3 000 m or −4 000 m on the stable margins and to −5 000 m or −10 000 m on the active margins. Its average slope of 4°–5° may seem to be gentle, but it nevertheless contrasts markedly with the slope of the continental shelf which is almost twenty times less (Fig. 1.3). The continental slope is dissected to varying degrees here and there by *submarine valleys or canyons* which can sometimes begin very close to the coastline.
  • On the stable continental margins the foot of the continental slope is occupied by the *continental rise* which slopes very gently between 1 per cent and 0.15 per cent

Fig. 1.3 The continental margin east of Argentina in the south Atlantic. The continental shelf is wide and the continental slope is cut by several canyons. The continental rise is bordered by an abyssal plain. Scale 1 cm = 80 km. (After Heezen and Tharp, 1961.)

generally from 4 000 m to 5 000 m deep. It is crossed by valleys or channels which diverge and fan out (Fig. 3.13). The rise can be situated on continental crust or oceanic crust.

● The active (or Pacific-type) continental margins have no continental rise. They are generally characterised by the presence of a *marginal trench*. These, on average, are between 70 km and 100 km wide and stretch for several hundred kilometres. It is here that the deepest parts of the globe are situated, such as the Tonga Trench, 10 km deep, and the Mariana Islands Trench, 11 km deep (Fig. 1.4).

(b) Ocean basins

The morphology of the *ocean basins* is determined to a large degree by the nature of the surrounding margins:

● The rises on the stable margins often abut on to very flat *abyssal plains* which are

3

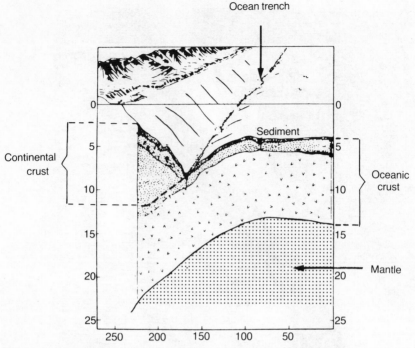

Fig. 1.4 A perspective view of the Chilean trench. Lithospheric structure from seismic refraction. Scale in km. (After Fisher, 1974, simplified.)

marked by a few *sea-mounts* and infrequent channels that extend from the valleys on the rise.

• The ocean floors bordering the active margins, on the other hand, have more relative relief. Apart from sea mounts which may rise several kilometres from the floor, bathymetric surveys have also revealed a multitude of *abyssal hills* which do not exceed a few hundred metres in height.

These hills also occur in fact beneath the plains at the foot of the rise, but they are buried by sediments and can only be brought to light by seismic reflection. In the case of the stable margins, the relief is buried by thick sedimentary deposits (especially by turbidites transported along the floor from the continent). In the case of the active margins the marginal trenches act as a trough barrier which prevents this sort of supply from reaching further into the oceans. Here the abyssal hills are only covered by deposits of pelagic origin from the slow sedimentation of particles transported in the open sea.

(c) Active mid-ocean ridges

The *active mid-ocean ridges* occupy a good third of the oceans (Fig. 1.5) and form vast masses rising from the sea-floors. Their bases are about 5 000 m deep and their summits about 2 500 m deep and they are several thousand kilometres wide. These ridges are considered active because they are marked by major seismic disturbances along their crests. They are found in all the oceans of the globe and form an assemblage which is 70 000 km long.

In detail these active ridges are broken up by a multitude of *transverse fractures* (transform faults) which offset each segment in relation to its neighbours. A *central*

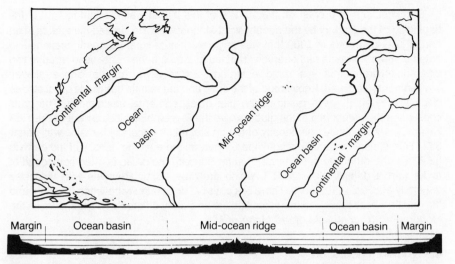

Fig. 1.5 The major morphological provinces of the north Atlantic. (After Heezen *et al*., 1959.)

valley is frequently observed along the symmetrical axis of the crest. This valley is only some 30 kilometres wide but reaches an average depth of 2 km. It is often called the 'oceanic rift' by analogy with the continental rifts with which it has certain characteristics in common.

(*d*) Other relief features

The ocean floors also display a disparate group of relief features of varying degrees of importance. These include *sea-mounts* which are most frequently old volcanoes, aseismic ridges (which are masses of great length without seismic activity) and volcanic islands which occur either singly or in archipelagoes.

Thus, even a very brief morphological description establishes an initial classification of the major structural features in the oceans. Active ridges and deep basins are in continuity and constitute the oceanic zone in its strictest sense. On the other hand, the continental shelf belongs to, and can only be separated with difficulty from, the continent itself. The deep continental margin is situated in the transition zone between the continent and the ocean. But a basic distinction should be made between the stable margins (where the transition can be described as the legacy of former features, since no tectonic activity occurs there at present), and the active margins which are, in contrast, the seat of present-day geodynamic features.

2 Continental crust and oceanic crust

In order to try and understand the deep structure of the globe and especially that of the lithosphere, geophysical methods such as gravimetry, seismology and magnetometry have to be used. Measurements of the velocities of seismic wave transmission (especially of longitudinal P-waves), enable estimates to be made of the density of the rocks crossed by earthquakes and also enable hypotheses to be advanced about the petrographic nature of these rocks.

5

The existence of a crust on the surface of this planet was brought to light at the beginning of the century by the geophysicist Mohorovičić. He studied the effects of an earthquake in Croatia in 1909 that was registered by several European stations. He noted that the P-waves had been divided and had thus been transmitted through two different environments – a 'rapid' environment (the upper mantle) and a 'slower' environment (the crust). Knowledge of the crust and the mantle has made great strides since then, notably as a result of the use of experimental seismology. The crust constitutes the outer shell of the globe where the P-wave velocities do not exceed 7.9 km/s. Correspondingly, the density of the crustal rocks generally remains well below 3.3. (This is also the density of the moon, whereas the average density of the earth is 5.52.) On the other hand, the *upper mantle* beneath the crust, is often composed of rocks with a density close to 3.3, which are crossed by P-waves with velocities generally exceeding 7.9 km/s. There is a physical discontinuity between the crust and the mantle where these seismic velocities are so clearly different. This is known as the *Mohorovičić discontinuity* or *the Moho* or *M*.

(a) The continental crust

On average the continental crust is 35 km thick. In some areas that figure can be twice as much. In general, the Moho is considerably deeper beneath young, high mountain chains where the crust can be 70 km–80 km thick. These are then called the *roots* of the chain. Conversely the crust tends to be much thinner on the continental rifts and more especially on the margins where the transition occurs between the continental crust and the much thinner oceanic crust (Fig. 1.6).

The continental crust not only varies in thickness. The distribution of 'seismic velocities' according to depth shows that it is also stratified. As a result of Conrad's study in 1923 on the effects of the Tauern earthquake, the distinction of two crustal layers has long been accepted:

• **The upper layer**, known as the *'granite'* layer because its physical properties are akin to those of granite or more correctly to those of *granodiorite* or *diorite* rocks. P-wave transmission velocities here are of the order of 6 km/s and the average density is 2.8. (The uppermost parts of this first layer can, however, be composed of less dense and less 'rapid' sedimentary rocks.)

• **The lower layer** is generally called *'basaltic'* for similar reasons (P-wave velocities between 6.5 and 7.7 km/s; densities between 2.9 and 3.1). But it is unlikely that this layer is actually composed of basalt or even gabbro. The pressure and temperature conditions at this depth must necessarily produce high-grade metamorphism giving rise to rocks such as *garnet-granulites, eclogites* and even to *amphibolites*.

(b) The oceanic crust (Table 1)

The oceanic crust differs markedly from the continental crust both in its average thickness and its nature. The oceanic crust is thinner (about 7 km, to which may be added the 5 km of the oceanic waters). The continental crust contains about 60 per cent SiO_2 whereas the ocean crust contains less than 50 per cent. The oceanic crust is thus denser and clearly has a more 'basic' petrographic composition.

• **Unconsolidated or poorly consolidated sediments** (layer 1) are only thin on average. (Half the oceanic sediments are preserved on the continental margins, i.e. beyond the ocean crust.) The thickness of layer 1 gradually increases from almost zero near the active ridge axes to an average of 500 m in the ocean basins (where thicknesses of 2 or 3 km may occur locally).

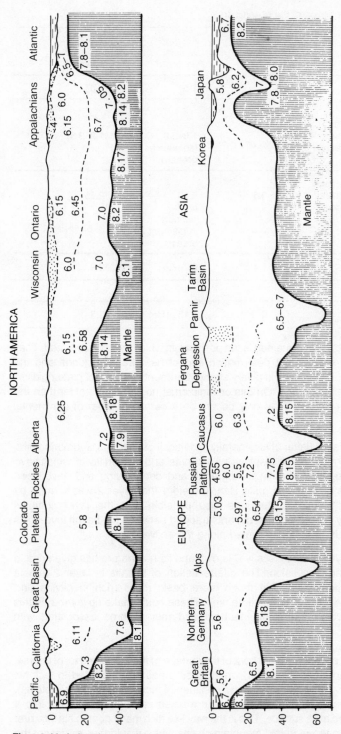

Fig. 1.3 Variations in crustal thickness in North America and Eurasia. The figures indicate P-wave velocities. Vertical scale in km. (After Holmes, 1965.)

7

Table 1   Average composition of the oceanic crust

	Average thickness (in km)	Average velocity (in km/s)	Average density
0 Sea water	4.8	1.5	1.03
1 Unconsolidated or poorly consolidated sediments (from lower Cretaceous to present)	0.45	1.8–2.0 (sometimes greater)	1.9–2.3
2 Basement: Volcanic rocks (pillow-lavas and basalt dykes) + consolidated sediments	1.7±0.6	5.10 (quite variable in different areas)	2.55
3 Oceanic layer	4.9±1.4	6.7–7.0	2.9
M Moho			
4 Upper mantle		8.13±0.24	3.3

All these sediments have remained in the positions in which they were formed. They have undergone no subsequent deformation except near fracture zones and the continental margins. (This in itself is a very strong argument in favour of plate rigidity.) They are relatively recent in age. On this oceanic crust no sediments older than the Jurassic are known (and the Trias only occurs on a certain number of continental margins).

● **The basement layer** (layer 2) is generally basaltic. It often forms 'submarine hills' on the surface and, in more detail, displays 'pillow lavas' and 'corded lavas' which have been observed through submarine photography and deep diving and have been studied through drilling. It seems quite clear therefore that layer 2 was formed by submarine volcanic eruptions. But it is also possible that the layer contains consolidated or partly consolidated sediments trapped between the basalt lava-flows when the oceanic crust was being formed along the active ridges.

● **The oceanic layer** is much less well known and its real nature has given rise to much controversy and discussion. The differentiation of basalts in layer 2 from a magma in the mantle implies that much more basic rocks (rich in olivine and plagioclase) would be formed at the same time. These could make up layer 3. In this case these rocks would be mainly gabbros and metagabbros associated with peridotites.

The oceanic crust occupies 60 per cent of the surface of the globe and is, of course, confined to the ocean areas. However, when these areas are subjected to tectonic stresses where plates converge (as in active continental margins and fold mountains), it can happen that parts of the oceanic crust are taken up and incorporated into a continent and exposed at its surface. This is the way in which many geologists interpret the famous *ophiolite nappes* in fold mountain chains which would thus represent the remains of former oceans which have disappeared through subduction.

(c) The 'intermediate crust'

The Moho discontinuity plunges markedly beneath the continental margins. In this transition zone, the crust acquires characteristics which are intermediate between the continental and the oceanic crusts. It is not, however, known exactly how the change from one type of crust to the other actually takes place. Nevertheless this is a major geophysical and geological problem which must be solved if stable continental margins (and especially their considerable subsidence) are to be fully understood. It is, however, precisely the great thickness of sediments that is preserved on these margins which has hitherto prevented drillings from reaching the substratum. Geologists and geophysicists must thus be content with hypotheses:

• P-wave velocities in the basement on stable margins are usually of the order of 6.5–6.8 km/s. This links this basement to the deeper part (or 'basaltic' layer) of the continental crust. This relationship can be interpreted by suggesting that the thinning of the continental crust occurred through the loss of its upper parts especially during the rifting before the margin was formed. In this case the intermediate crust would be a thinned continental crust partly composed of high-grade metamorphic rocks.

• Stable continental margins form as a result of lithospheric extension. Apart from crustal thinning resulting from the development of normal faults, the injection of dense volcanic rocks from the mantle can also occur. The intermediate crust would be in this case a continental crust which had been disorganised and enriched by basic magma intrusions.

This discussion will be resumed and developed in Chapter 2.

3 Lithosphere and asthenosphere

The *lithosphere* is the outer shell of the globe and includes the crust and part of the upper mantle. It is characterised by both *rigidity* and *elasticity*. In more precise terms, the constituent rocks can withstand stresses of the order of 1 kilobar without flowing. The lithosphere has an average thickness of about 100 km. It is covered by the hydrosphere and the atmosphere and, in turn, lies upon the asthenosphere.

The *asthenosphere* on the other hand is composed of *plastic* materials which are deformed when subject to weak stresses of the order of about 100 bars. More graphically, the asthenosphere can be seen as a viscous milieu (at least on the scale of geological phenomena) which not only allows the lithosphere to move across the surface of the globe but enables it to plunge several hundreds of kilometres below this surface.

(a) The bases of the concept of the lithosphere and the asthenosphere

The ideas about the asthenosphere and the lithosphere which are accepted today have gradually appeared during the present century. They are based, in particular, on gravimetric, seismological and geothermal observations:

• When a great weight is applied to the surface of the globe (where for example a delta or a volcano has been built up), a regional deformation of this surface may be observed (Fig. 1.7). The lowered zone is much more extensive than that to which the weight is directly applied and, beyond it, the lowering is compensated for by a broad arching. Regional isostatic equilibrium is maintained, but a marked positive gravimetric anomaly may be observed locally where the weight has been applied. Such features

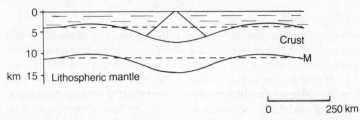

Fig. 1.7 The effects of a load (here the Hawaiian archipelago), applied to the lithosphere. (After Coulomb and Jobert, 1973.)

can only be explained by the presence of a rigid and elastic milieu (the lithosphere) resting upon a viscous milieu (the asthenosphere) which can flow and thus re-establish isostatic equilibrium by allowing materials to migrate at depth.

● Seismic waves (longitudinal P-waves and more clearly transverse shear S-waves, which are slower than P-waves) have velocities depending on their depth. These waves are fairly rapid at the surface but are slowed down in a 'low velocity zone' some 100–300 km below the surface. These variations in velocity imply that the physical properties of the upper mantle rocks change along a horizon at a depth of about 100 km which may be defined as the lithosphere–asthenosphere boundary.

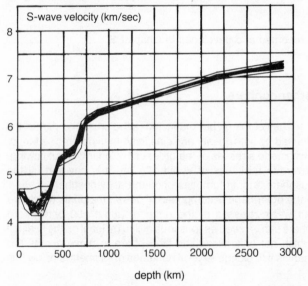

Fig. 1.8 Distribution of S-wave transmission velocities according to depth. (After Press, 1970.)

● Knowing the average value of the heat flow at the surface of the globe and also the nature of the rocks in the upper mantle, it is possible to calculate the depth of the 'solidus' of these rocks, or the point where their partial melting begins. This method indicates a depth of some 75–300 km. It seems therefore that the base of the lithosphere is marked by an isotherm (of about 1 100° C). It also seems that the slowing down of S-waves and the plasticity of the asthenosphere are both caused by the effects of the start of partial melting. This melting probably involves but a small part of the mantle material (~ 1%), but nevertheless it is enough to modify its physical properties to a considerable degree.

Thus three relatively independent methods point to the same conclusions. The mantle clearly possesses a rheological stratification which is essentially linked to a thermal structure. This stratification can however vary laterally. While the lithosphere is about 75–80 km thick in the oceans, it increases to 120–130 km beneath the continents. It even seems that the 'low velocity zone', which can always be clearly identified beneath the oceans, is less well developed under the Ancient Shields. Thus the contrast between the ocean and the continent which is clearly evident when the two types of crust are compared, also occurs in the lithosphere.

(b) The lithospheric mantle

The lithospheric mantle beneath the Moho discontinuity has a density of 3.3 and transmits P-waves usually with a velocity of 8.1 km/s. It is now acknowledged that it is composed of peridotites.

The lithospheric mantle is not homogeneous. Recent experimental seismological studies, especially off western Europe have brought a clear stratification to light. Just under the Moho an initial layer 10 km thick transmits P-waves at an average velocity of 8.1 km/s. Then a 'low velocity channel' about 10 km thick is observed in which the P-waves are transmitted at 7.8–7.9 km/s. Finally the base of the lithosphere becomes 'rapid' once again with velocities of 8.2 km/s (Fig. 1.9).

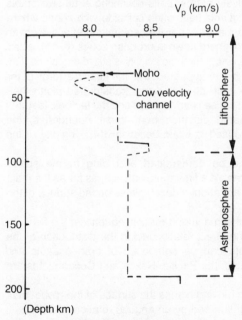

Fig. 1.9 Velocity-depth relationships beneath the western European crust (P-wave transmission). (After Hirn, 1976.)

This upper low velocity channel clearly cannot be interpreted like the asthenosphere as the result of the partial melting of mantle rocks. On the other hand, some authorities believe that it could be caused by the hydration of the peridotites. The serpentinisation of ultramafic rocks ought normally to occur at the pressures and temperatures pertaining at a depth of 40–50 km, provided the environment contains a small proportion of water (1%).

The discovery of this low velocity channel in the lithospheric mantle is undoubtedly of considerable geological interest. If a thick layer (of about 10 km) of serpentinised peridotite divides the lithosphere horizontally, it is highly likely that any tangential tectonic force would cause decoupling along it. This feature could explain the frequent association of high-grade metamorphites, which have been crystallised at the base of the continental crust, and peridotites (Cherzolites) in many fold mountain chains (in the Ivrea zone in the Alps, the north Pyrenean zone, and the Alpujarrides in the Betic Cordillera, etc.). It may also explain the presence of peridotites at the same time as rocks attributed to the oceanic crust in the great ophiolite nappes. In both cases, these would be deep remnants of the upper lithosphere that have been decoupled from its lower parts along the serpentised layer and carried up to the surface by tectonic action and then exposed by erosion. (Ch. 6)

4 Plate tectonics

In the preceding paragraphs the lithosphere was considered as a continuous shell. In fact it is broken into a certain number of rigid spherical fragments called *'Plates'*.

The plate concept grew out of observations of the surface seismicity of the globe. Earthquake epicentres are not scattered haphazardly over the continents and oceans, but instead form continuous and relatively narrow belts delimiting extensive stable surfaces (Fig. 1.10). Moreover these seismic belts often coincide with zones where other geodynamic features occur: the formation of new lithosphere along active ocean ridges (p. 14), the disappearance of lithosphere down subduction zones (p. 15), active deformation along fold mountain chains where tectonic activity is still taking place (Ch. 6), etc. Thus the basic principle of the theory appeared: that the kinetic energy of the plates which are moving in relation to each other is partly transformed along their boundaries. Here the friction occurs that causes earthquakes and the tectonic features studied by geologists. Elsewhere, some distance from these boundaries, the geodynamic features are markedly reduced in scale because of the rigidity of the lithosphere.

Conversely, plate boundaries must be demarcated according to the spatial distribution of these geodynamic features. At a first approximation, as far as the main structural groups are concerned, there are about a dozen plates on the surface of the globe at present. (Fig. 1.10)

These principles of the theory entail some important consequences:
• The configuration of the plates has no direct relationship to the distribution of the oceans and the continents. One plate can be composed of both oceanic and continental lithosphere. Only three plates, the Pacific, Nazca and Cocos plates are wholly oceanic.
  • As the plates are spherical shields moving across the surface of the globe, their motion can be defined by a pole of rotation and by an angular rotation velocity.
  • Three sorts of relative motion between plates are possible: separation (divergence), closing together (convergence) and friction (shearing) (Fig. 1.11).

(a) The effects of plate divergence (Fig. 1.12)

The idea of sea-floor spreading preceded the plate theory by a few years. In order to explain Continental Drift and the distribution of magnetic anomalies in the ocean floors, geologists and geophysicists considered that the axes of active mid-ocean ridges marked the place where oceanic lithosphere and crust were continually being formed.

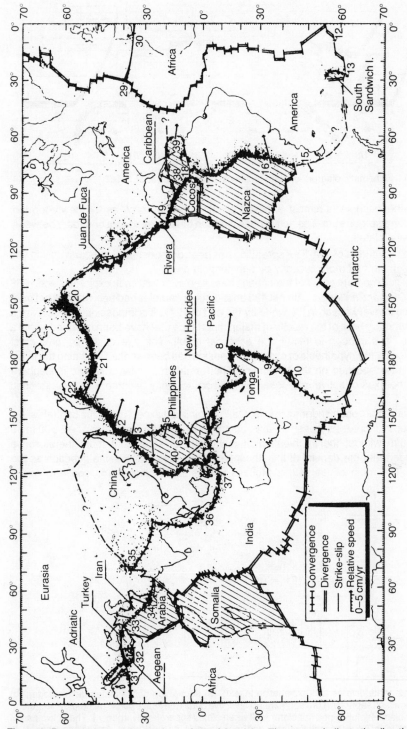

Fig. 1.10 Present plate motion on the surface of the globe. The vectors indicate the direction and velocity of relative motion. Epicentres of major earthquakes are represented by dots. (After Le Pichon *et al.*, 1973.)

13

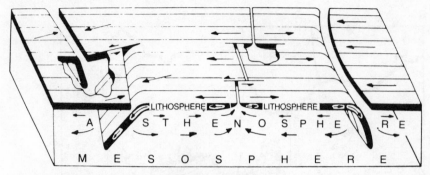

Fig. 1.11 Schematic diagram of plate tectonics. (After Isacks *et al.*, 1968.)

The asthenosphere is almost exposed on the sea-floor there as the newly-formed lithosphere spreads out. The active ridge axis can thus be seen as a boundary between 'divergent' plates.

The hypothesis of sea-floor spreading has been demonstrated by ocean drilling and has been verified more recently by submersible dives.

The sedimentary cover of the oceans has been systematically explored since 1968 by the *Glomar Challenger*. Almost 430 drills for core samples had been carried out from this ship in every ocean in the world by the end of 1977. It is impossible to give even a brief summary here of the results of major importance that have been obtained within a decade. It is enough to recall one essential result. The age of the oceanic crust predicted by the hypothesis of oceanic expansion has been verified by dating the initial sediments deposited on the basement. The formation of new crust and lithosphere along the axes of active mid-ocean ridges consequently became a well-established fact.

● Active mid-ocean ridges are in isostatic equilibrium except along the central valley. The depth increases from the axis down to the ocean basin from $-2\,500$ m to $-5\,000$ m without the intervention of any tectonic features. It must therefore be concluded that the density of the lithosphere increases in the same direction as the depth. The main cause of this process seems to be *lithospheric cooling*. Young

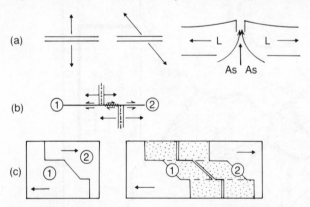

Fig. 1.12 Plate divergence (explanation in text). (a) separation of two plates (active ridge); (b) transform fault separating two segments of an active ridge (the active part is marked by hachures); (c) origin and evolution of a transform fault when plates separate (divergence). The active part of the fault is in full lines; the inactive and fossil part in dashed lines. As: asthenosphere; L: lithosphere.

lithosphere near the crest of the ridge is 'warmer' than old lithosphere which forms the floors of the ocean basins. In other words, the gradual lowering of the relief on either side of the ridge crests is caused by the ageing of the newly-formed lithosphere which gradually reaches its thermal equilibrium. Heat flow studies on the ridges confirm this interpretation. (Ch. 2).

• The thickness of the sedimentary layer (layer 1 of the ocean crust) increases towards the ocean basins. It may be zero on the ridge crests but may reach a thickness of about 1 km at depths exceeding −4 500 m or −5 000 m. The hypothesis of sea-floor spreading explains this arrangement by the increasing age of the crust. The crust is covered by a sedimentary layer which is thicker where the time available for deposition has been longer − i.e. where the ocean basement is older.

• The basalt magma cools along the axes of the active mid-ocean ridges and acquires a thermoremanent magnetism when its temperature falls below the Curie point. Moreover, the earth's magnetic field has been frequently reversed during geological time. It follows that the magnetism of the basalt layer must change its polarity according to the period when it was formed. Magnetic contrasts are thus formed in the crust which cause the anomalies in the intensity of the magnetic field observed in the ocean. These magnetic anomalies thus represent kinds of 'growth stripes' in the ocean crust. The older the stripes, the further they are found from the ridge axis.

• The ridge axes correspond to a narrow band of tectonic activity where earthquake foci are 'shallow' (i.e. occur at depths of less than 70–100 km, which is the average thickness of the oceanic lithosphere).

• The relief of the mid-ocean ridges differs according to whether the ocean has expanded at a 'slow' or 'rapid' rate (more than 3 cm per year). In the first case a central valley occurs, whilst in the second case, the ridge forms a wide, round-topped height without an axial trench.

• Finally, divergent plate boundaries also mark the place where newly-formed lithosphere interacts with sea-water. The real importance of such a feature was not suspected until recent years. It is now known, not only that the alteration and metamorphism of the ocean crust and the lithospheric mantle result from hydrothermal processes caused by sea-water infiltration, but also that sea-water itself takes up salt during its 'journey' in the warm lithosphere before returning to the ocean. This helps to explain the salinity of sea-water which is not only a legacy of the distant geological past but also results from a permanent exchange with the lithospheric crust and mantle.

(b) The effects of plate convergence (Fig. 1.14)

If the plates can move apart at speeds of several centimetres per year, then either the globe is rapidly expanding, or the creation of new lithosphere at the mid-ocean ridges is matched by an equal amount of lithospheric destruction in other parts of the globe. It seems, however, that the volume of the earth has undergone practically no variation in the last 200 million years, i.e. the period during which the present oceans were formed. Plate divergence therefore necessarily implies an opposing convergent motion, which occurs on the active continental margins (Chs. 4 and 5).

• This second type of (convergent) plate boundary is marked on the earth's surface by a seismic belt which differs considerably from that along the axes of the active mid-ocean ridges:
 (i) A much greater amount of energy is dissipated (Fig. 1.10) and the seismic belt on the active margins is much more clearly marked than the belt in the ocean rifts.
 (ii) The earthquakes are not only shallow (0–100 km), but are also intermediate or deep (at depths of −100 km to −700 km).

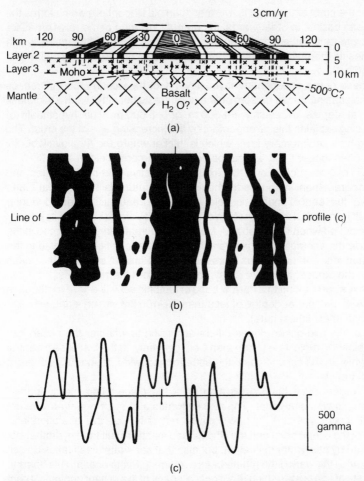

Fig. 1.13 Explanatory diagram of magnetic anomalies according to the hypothesis of sea-floor spreading as formulated by Vine and Matthews. In black, normally magnetised material; in white, reversed; (a) evolutionary diagram; (b) part of the corresponding map of magnetic anomalies; (c) magnetic profile seen on the line shown in (b). Expansion rate: 3 cm/year. (After Vine, 1968, quoted by Le Pichon, 1973).

(iii) The geometric position of these earthquake foci forms a markedly flattened zone which has been given the name 'Wadati–Benioff zone'.[1] This zone dips downwards and its intersection with the surface of the globe occurs along the ocean trenches on the active margins (Fig. 1.14(a)).

• The means by which a plate is thrust beneath another plate into the asthenosphere is called *subduction*.

In principle only oceanic lithosphere can penetrate into the asthenospheric milieu. It will be recalled that the oceanic lithosphere emanates from the asthenosphere but it has a higher density because it is cooler. It can therefore sink without difficulty into a less dense and more viscous milieu. In contrast, the continental lithosphere (with its thick and relatively light crust) has a lower density than the asthenosphere. This normally prevents it from being involved in subduction. Consequently, plate convergence should stop as soon as a part of the lithosphere carrying some continental crust reaches an

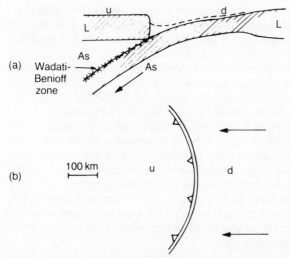

Fig. 1.14 Convergence and subduction. L: lithosphere; As: asthenosphere; u: over-riding plate; d: descending plate. (a) diagram in cross-section; (b) diagram in plan.

ocean trench. Intense deformation of the continental margins thus brought into contact may then be expected and they become elements in a fold mountain chain. This phenomenon is called *collision* (Ch. 6).

• At the surface of the globe the boundaries of convergent plates usually form arcs whose concave edge is situated on the same side as the over-riding plate (Fig. 1.14(b)). Fracture zones (transform faults) are often observed between two arcs where two plates are moving past each other (Fig. 1.15).

Fig. 1.15 Examples of arc-to-arc transform faults.

(c) Transform faults

Transform faults constitute the third family of lithospheric boundaries. They have been introduced already in the two foregoing paragraphs. They are fracture zones where plates basically slide past each other (dextrally or sinistrally) (Fig. 1.12(b) and (c); Fig. 1.15). This causes some important shallow seismicity. By definition these transform faults coincide with plate boundaries along which the lithospheric surface is neither being destroyed nor created. There are three major categories of transform faults:
 (i) rift to rift faults
 (ii) rift to arc faults
(iii) arc to arc faults.

• The concept of transform faults must be clearly separated from that of transcurrent faults, with which geologists are more familiar. Transform faults suddenly stop being active at the ends of the arcs or rifts which they link together. They do not continue beyond the arcs. On the other hand, they extend beyond the rifts in the form of inactive fossil fractures which can be followed right up to the continent–ocean boundary (Fig. 1.12). The classic problem of where the great transcurrent faults, described by geologists, end thus becomes irrelevant in the case of transform faults. The sliding motion does not 'end': it is 'transformed' into another movement.

• The transform faults, whether they be active or fossil, are often very clearly marked in the relief, especially where the fracture zones run from rift to rift. The sliding motion causes the formation of horsts and fault-troughs which can reach depths of 6–7 km (e.g. the Romanche trench in the equatorial Atlantic which is 7 631 m deep).

• Rift to rift transform faults can be considered, at least in part, to be a legacy of the geometry of the original ocean (Fig. 1.12(c)). When a plate divides into two, the initial break is rarely straight. It usually occurs along zones of weakness that correspond to old geological structures whose surface expression can be most irregular. As the oceanic rift generally occupies the midst of the ocean that is being formed, its outline reproduces the line of the initial break fairly closely. If certain segments of this break trend parallel to the direction taken by the moving plates when the ocean begins to open, they give rise to transform faults which last as long as the rotation poles of the divergent plates remain approximately stable.

Notes

1. The real 'inventor' of this seismic zone was the Japanese geophysicist Wadati. But it has become common to call this zone after the American geophysicist, Benioff, who studied it more recently.

Further reading

Francheteau J. (ed.) 1979. 'Processes at mid-ocean ridges', *Tectonophysics* **55** (Special Issue), 260 pp.
Fuchs K. and **Bott M. H. P.** (eds) 1979. 'Structure and compositional variations of the lithosphere and asthenosphere', *Tectonophysics* **56**, 1–2 (Special issue), 201 pp.
Le Pichon X., Francheteau J. and **Bonnin J.**, 1973. Plate Tectonics. *Development in Geotectonics*, **6**, Elsevier Publishing Company, Amsterdam, 300 pp.
Talwani M., Harrison C. G. and **Hayes D. E.** (eds) 1979. 'Deep drilling results in the Atlantic Ocean: Ocean crust' *Maurice Ewing Series* 2, Amer. Geophys. Union, Washington DC, 437 pp.

Chapter 2
The geodynamic causes of the subsidence of stable margins

The stable margins are the sites of probably the greatest subsidence on the face of the earth. In this chapter, the various geodynamic factors causing this subsidence are reviewed and their effects are compared: crustal extension in the continental rifts which are themselves stable margins in embryo (§1); subaerial erosion in these rifts and replacement of the eroded parts of the crust by sediments during subsequent lithospheric cooling (§2 and §3); sediment load (§4) and any metamorphism in the deep crust (§5); and finally, continental crustal thinning through creep or extension which ends in the formation of intermediate crust (§6).

The sedimentary layer on the surface of the crust usually forms a thin film both on the continents and in the oceans (Ch. 1). In contrast, it increases considerably in thickness (and locally may exceed 10 km) on the present-day continental margins or in the fossil margins represented by the fold mountain chains. More than half the marine sediments have thus been deposited on the continental borders and more especially within the sedimentary basins of the 'Atlantic' type margins.[1]

The main geodynamic phenomenon dominating the evolution of stable margins is subsidence. Two kinds of margin may be distinguished according to the thickness of the sedimentary layer:
(i) 'Starved' margins, which are 2–4 km thick beneath the shelf and average 4 km in thickness beneath the continental rise.
(ii) 'Nourished' margins, which are 5–12 km thick on the shelf and 3–6 km thick on the continental rise.

In the Atlantic Ocean, the starved margins occur on the eastern side (and notably on the European margins), whereas the greatest thickness of sediments occurs on the western side along the American coast (Fig. 2.1). However, the stable margins are not only lowered under the weight of their accumulated sediments, they are also, indeed primarily, a zone of active subsidence which is generated from within the lithosphere. Thus, despite their names, these 'stable' margins form one of the relatively mobile belts on the surface of the globe.

1 The effects of lithospheric extension in continental rifts

The stable continental margins are old plate edges. The ocean–continent transition zone forms right at the start when two plates carrying continental crust diverge and the initial narrow band of oceanic lithosphere takes its place between them. Even earlier

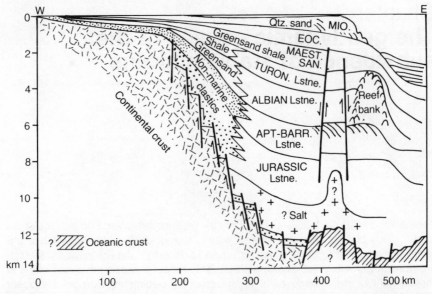

Fig. 2.1 A 'nourished' continental margin: the Atlantic margin of the United States. (After Sheridan, 1976.)

than this, before any oceanic growth occurs, the extension of the continental crust causes the appearance of a special type of structure, the 'continental rift', which is really a stable margin in embryo (Fig. 2.2). Thus, in order to interpret the evolution of Atlantic-type margins correctly, it is essential to understand the geodynamic features which may be observed in the rifts in East Africa (in the Rift Valley and in Afar), or in France (in the Limagnes of the Massif Central and in the Vosges and Rhine Rift Valley in Alsace).

The continental rifts usually form relief features on the earth's surface. The continental lithosphere is 'heated up' there, as can be observed in the heat flow anomalies which may locally reach two or three times their normal value. As a result the density of the lithosphere is lowered and a regional uplift takes place which varies in extent from place to place:

• The initial stage in the evolution (Fig. 2.2) seems to be a simple elliptical arching. On average, it is 1 km high, 100 km wide, 200–300 km long and has a surface area of 10^4 to 10^5 square km.

• Then (or at the same time), the uppermost part of the dome collapses. The classic fault-trough appears which divides the rift into two symmetrical parts. On average, the ridgecrests are 1.5 km high, the graben is 30–100 km wide, whilst the rift as a whole (including its flanks) is 200–2 000 km wide, and between several hundred and several thousand kilometres long. The relief of a continental rift at this stage closely resembles that of a mid-ocean ridge.

• Finally, the continental rift, which has hitherto been formed by a localised crustal extension, can be transformed into a zone of oceanic opening. The rift axis then becomes the boundary between diverging plates where lithosphere begins to form. The sea, which had previously penetrated only intermittently into the fault-trough, floods in permanently. It gradually increases in depth until it reaches the usual level of the ocean-floors. (The Afar region is probably exceptional in that oceanic crust is being formed there in the open air; Fig. 2.2.)

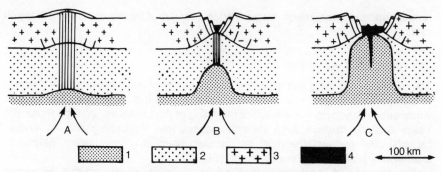

Fig. 2.2 Interpretative diagram of the various stages in the tectonic and magmatic evolution of the Afar region. 1: asthenosphere and low velocity zone; 2: upper mantle; 3: continental crust; 4: oceanic crust.
Stage A (64–25 m.y.) Formation of Arabian–Nubian arch. Thermal expansion of mantle and eruption of (mainly alkaline) basalts formed by partial melting at depth. This phase should be related to the Jurassic phase in the Atlantic.
Stage B (25–4 m.y.) Formation of the Afar, Red Sea and East African continental rifts.
Stage C (4–1 m.y.) Separation of the continental plates and formation of oceanic crust. Shallow partial melting in the mantle.
(After Treuil and Varet, 1973, simplified.)

This chronological evolution can also be observed spatially. In East Africa, for example, the continental rift system marked by the great lakes and the Afar region passes into zones of submarine oceanic growth in the Red Sea and in the Gulf of Aden.

(a) The deep structure

The continental rifts have been the subject of much experimental seismological work during the last few years and consequently their deep structure is fairly well known. They all display a series of common characteristics which are fairly clear in all the examples studied in spite of local differences.

• The continental rifts are characterised by weak but fairly constant seismic activity. The earthquake foci are shallow (0–30 or 40 km and are rarely deeper). Focal mechanism studies have clearly shown that the tremors are prompted in nearly every case by regional extension producing normal faulting. The displacement of the two edges of the rift is, however, still very restricted (1–5 km at the most) and also very slow (a fraction of a millimetre per year). Thus, in general the axes of continental rifts are not classed as plate boundaries.

• The continental crust becomes thinner near the axis of the rift. For example, the Moho rises by 5 km in the southern part of the Rhine Rift Valley where almost 3 km of Tertiary deposits have accumulated. The basement is thus 8 km thinner here. In the same way, the Moho rises to within 24 km of the surface in the Limagne graben, which contains about 2 km of Tertiary deposits, whilst the crust is 30 km thick in the Massif Central.

• In the axes of the continental rifts, the upper mantle often displays abnormally low seismic velocities, of about 7.6–7.4 km/s, which are usually explained by a local rise in temperature and by magma intrusions. But this does not always occur. It seems, for example, that the P-waves are transmitted at 8.1 km/s in the upper mantle beneath the Rhine Rift Valley. This is a 'normal' velocity.

• Finally, the fault-troughs in the rifts are marked by a negative Bouguer anomaly which reaches several tens of mgals (−30 mgals, for example, in the Rhine Rift Valley). The bulk of this anomaly seems to be caused by the sedimentary infill in the basin which

21

remains approximately in isostatic equilibrium because of the rise of the Moho in the rift axis.

(b) Associated vulcanicity

The extension occurring on the continental rifts is accompanied by intensive volcanic and plutonic activity. There are many examples of this in France. The famous volcanoes in Auvergne are a result of the rifting of the Limagnes, and the Rhine Rift Valley similarly displays large volcanic masses (the Vogelsberg and Kaiserstuhl Massifs). But perhaps the most spectacular example of magmatism related to extension is provided by the Afar region (Fig. 2.2).

Petrological and geochemical studies have clearly demonstrated that the source region of the magma is situated beneath the crust. The evolution of the magma, which is basalt initially, can be explained essentially by fractional and partial melting processes in the upper mantle (notably at the 'abnormal mantle cushion'), and then by fractional crystallisation within the magma. Thus rocks are created which are most often alkaline in nature and form a continuous series from basalts (which are by far the most abundant) to rhyolites.

The continental rifts are therefore areas where the continental crust is enriched by intrusions (and extrusions) emanating from the mantle. The density of the crust can thus be increased, which necessarily entails an isostatic reaction prompting crustal subsidence. Finally, magmatic intrusions often give rise to the magnetic anomalies observed on the surface which persist throughout geological history and which can be found especially on the stable continental margins.

(c) The fracture pattern and the formation of the fault-trough

The fault-troughs in the midst of the continental rifts are delimited by a network of normal faults (Fig. 2.3). This network has a very complex spatial distribution (Fig. 2.4). However, the faults trending parallel to the axis of the trough usually have the greatest throw and they cause the lowering and tilting of the staircase-like arrangement of crustal blocks. The central graben is often divided into two. In France, for example, the Loire and Allier (Limagne) troughs in the Massif Central are separated by the horst forming the Monts du Forez.

Figure 2.5 illustrates a model of fault-trough evolution. After an initial arching (Fig. 2.5(a)), the group of faults f_1 appears which cause the formation of a half-graben. Then, as extension continues, a second group of normal faults f_2 begins to come into play. The rift fault-trough is thus formed which is overlooked by two ridges that undergo vigorous erosion (§3). However, evidence of the initial (half-graben) phase of evolution is inscribed in the asymmetrical sedimentary infill, which has an older and thicker side (f_1) and a more recent and thinner edge (f_2).

In detail, the groups of faults f_1 and f_2 (which will belong to two opposite margins if the rift develops into a zone of oceanic accretion) separate a complex assemblage of tilted blocks. In the graben these form small second-order fault-troughs, or (more frequently) monoclinal troughs whose lowered edge is usually situated on the side away from the axis of the rift (Fig. 2.6). These structures are very familiar to the geologists studying continental margins and they are clearly a legacy of the initial continental rift.

In summary, the continental rifts correspond to zones where the lithosphere undergoes limited extension. Here several kilometres of crustal thinning occur and fault-troughs develop with thick sedimentary accumulations. The continental crust, moreover, is enriched by a multitude of magma intrusions and extrusions emanating from the mantle. If a rift is changed into a zone of oceanic accretion, its structures and sediments are incorporated into two opposing stable continental margins (*cf.* (a) p. 37).

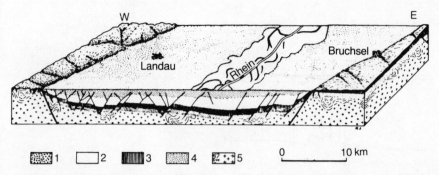

Fig. 2.3 The Rhine graben north of Karlsruhe. 1: Neogene and Pleistocene; 2: Palaeogene; 3: Middle and Upper Trias and Jurassic; 4: Permian and Lower Trias; 5: Hercynian basement. (After Illies, 1974.)

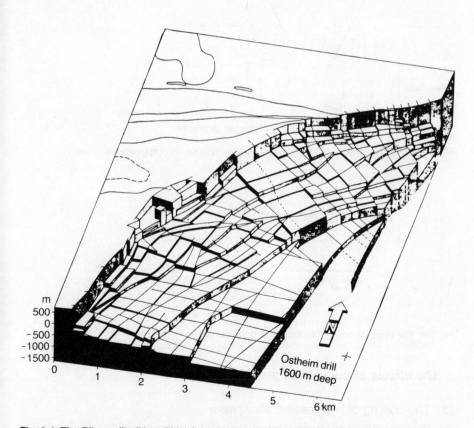

Fig. 2.4 The Ribeauville (Haut Rhin) fracture system. The wall of Vosges sandstone acts as a reference surface. (After Hirlemann, 1974.)

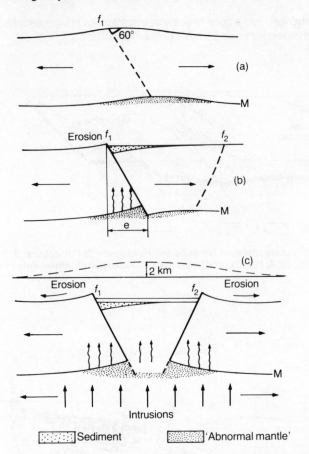

Fig. 2.5 Schematic model of the evolution of a continental rift (explanations in text.)

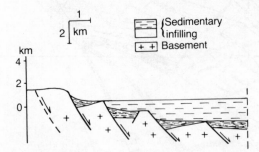

Fig. 2.6 The tilting of crustal blocks in a continental rift.

2 The effects of cooling on the lithosphere

(a) The cooling of the oceanic lithosphere

On either side of the axis of an active mid-ocean ridge, the oceanic lithosphere deepens as it increases in age (Ch. 1). This can be studied with some precision because both the depth of the ocean and the age of the crust are known through deep drilling, seismic studies and the distribution of magnetic anomalies (Fig. 2.7). Broadly speaking, the depth increases with age exponentially with a time constant of the order of 50 million years.[2] The average depth on the axes of the mid-ocean ridges is about 2 550 m if the expansion is slow and 2 700 m if it is rapid. In 80 million years, the ocean floor is lowered by about 3 000 m. Conversely, it is possible to discover the approximate age of the crust if its depth is known.

The curves in Fig. 2.7 represent 'corrected' values. In order to eliminate the variations in depth caused by localised changes in sediment thickness (layer 1), the position of the basement was calculated assuming that layer 1 has been removed and the corresponding isostatic correction has been made. But, the weight of the sea-water (density 1.05 per cm[3]) must also be taken into account, if the real effect of lithospheric contraction is to be envisaged properly. The 3 000 m of lowering reached in 80 million years should therefore be reduced to about 2 000 m.

The main cause of this subsidence seems to be the cooling of the newly formed lithosphere which progressively increases its density. Heat flow measurements in the central valley and on the crests of the mid-ocean ridges reveal a marked positive anomaly (3 or 4 μcal/cm²/s or H.F.U. and occasionally more). In contrast, the flow on

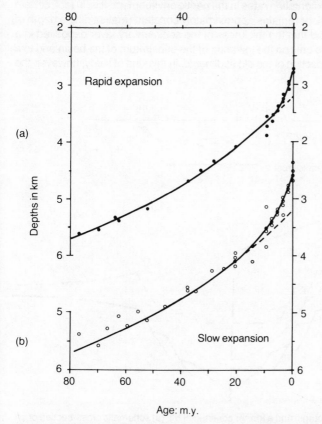

Fig. 2.7 The depth of the oceanic basement as a function of the age of the crust. (It has been assumed that the sedimentary layer has been subtracted and the corresponding isostatic correction has been made). (a) rapid oceanic expansion (>3 cm/year); (b): slow oceanic expansion (< 3 cm/year). (After Sclater *et al.*, 1971, synthesised by Le Pichon *et al.*, 1973.)

the flanks decreases as the depth increases and as the crust ages. Abnormally low values are thus registered in the oldest basins (0.9–1 H.F.U.). Thus a supplementary heat loss from the newly-formed lithosphere should probably be added to the earth's normal heat radiation. The newly-formed lithosphere itself only reaches thermal equilibrium after several tens of millions of years.

(b) The cooling of the continental lithosphere

The continental lithosphere is clearly different from the oceanic lithosphere (Ch. 1), and the laws of thermal subsidence established in the oceanic environment can only be applied to the continent with some caution.

The cooling of the continental lithosphere occurs in two ways:
(i) when a continental rift ceases to be active (it is then said to 'abort' because it does not give birth to an ocean)
(ii) when, on the other hand, continued ocean expansion causes the seaward migration of the high heat flow zone (the mid-ocean ridge), and a consequent cooling of the initial continental rift which thus becomes the margin of the developing ocean (Fig. 2.8).

In both cases the means by which the lithosphere was lowered can be reconstructed from variations in the sedimentation rates in the neritic environment. It is, in fact, certain in this case that the sea-floor remains approximately constant (maintaining a depth of between 0 and 200 m) and that the thickness of the sedimentary layer deposited in a given unit of time depends only on the lowering of the substratum of the basin and to a lesser degree on the compaction of the old sediments. In this kind of study, however, the

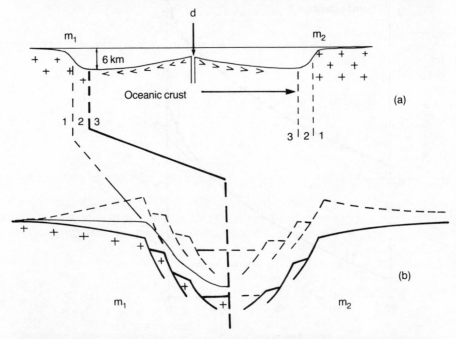

Fig. 2.8 Stable continental margin and a former continental rift. (a) schematic cross-section of an ocean bordered by two stable margins m_1 and m_2 where oceanic accretion occurs on an active mid-ocean ridge d; (b) detail of figure (a). The dashed lines represent the former continental rift that has been lowered by thermal contraction. 1: former flanks of rift; 2: former graben; 3: zone of newly formed oceanic crust.

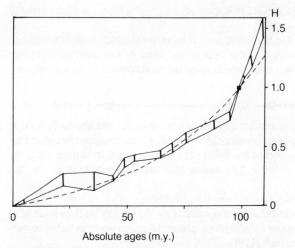

Fig. 2.9 Depth to which sediments are buried as a function of their age on the continental shelf of the eastern United States (H = $\frac{h}{ho}$, where h is the depth to which beds of a given age are buried, and ho is the depth of the end-Cretaceous where the drilling took place). The curve in a dashed line is the curve of an exponential subsidence that is identical to the mid-ocean ridge. The two full-line curves enclose 90 per cent of the points of observation. (After Sleep, 1971.)

fault-trough zone must be avoided because the subsidence there can have other causes than thermal contraction (*cf.* §1, p. 19). Lastly, the laws of subsidence caused by continental lithospheric cooling have been established either in intracratonic sedimentary basins or on the continental shelves on the flanks of the old rifts. The variations in sedimentation rates are known there from the considerable number of industrial drillings undertaken in these areas.

The most important result of this research has been the demonstration that the thermal subsidence of the continental lithosphere obeys the same laws as the oceanic lithosphere in the majority of cases and especially on many continental shelves. Lowering in relation to time is very close to an exponential curve with a time constant of some 50 m.y. (Fig. 2.9). In the evolution of stable margins, then, there is no need to consider the cooling of the continental and oceanic lithospheres separately. In spite of their markedly different origins, both are found in a fairly similar thermal condition as the margin starts to develop. They then cool at the same time and cause comparable amounts of subsidence.

Conversely the similarity of their behaviour during cooling explains why the warming of both kinds of lithosphere has the same morphological effects. In particular, the way in which the ocean basement lowers from the 'warmest' zone on the ocean ridge crests to the 'coolest' zone in the deep ocean basins, which is about 2 000 m (after hydrostatic and isostatic corrections have been made), may be compared with the average height of the continental rift crests (about 1 500 m) which have, however, undergone erosional attack.

3 The effects of subaerial erosion on continental rifts

The resemblance in form between a continental rift and a mid-ocean ridge, however, should not be allowed to hide an essential difference. The mid-ocean ridge evolves beneath the sea and its basement is scarcely eroded at all. In contrast, the continental rift forms a relief feature that is subject to subaerial action which brings about crustal thinning by the erosion of its surface.

This erosion is not restricted to the 1 500 m of relative relief which characterises an active continental rift. The removal of the uppermost parts of the crust prompts an isostatic response and a further uplift which may be represented schematically as follows:

(thermal) uplift ⟶ erosion ⟶ (isostatic) uplift

It is thus possible to calculate the amount of crust eroded from the lips and flanks of the rift. Given that the rift is active over a considerable period of time, erosion should, in the end, cause a progressive lowering of the relief until a 'peneplain' is formed close to sea-level. The crustal slice x $(= x^1 + 1.5)$ that is thus eroded away is given by the equation (Fig. 2.10):

[1] $(1.5 + x^1)\ 2.8 = x^1 \times 3.15$ ⟶ $x = 13.5$ km

which respects the isostatic equilibrium in the initial state (b) and in the final state (c). (The densities chosen for the upper mantle (3.15 g/cm³) take into account its particular thermal state beneath the continental rift.)

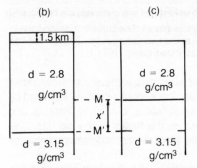

Fig. 2.10 Erosion on the lips of a rift in long-continued activity.

The hypothesis whereby the relief features are completely destroyed is not, of course, realistic. But a continental rift can remain active for very long periods. The Rhine Rift Valley, for example, was created at the end of the Cretaceous. In such conditions, the removal of 5 km of continental crust does not seem unlikely.

When such a rift stops being active either because it is 'aborted' or because it is transformed into a stable margin, it is lowered under the influence of lithospheric contraction:

(thermal) subsidence ⟶ sedimentation ⟶ (isostatic) subsidence

The thickness y of the sedimentary layer accumulating in this way depends on the quantity x of the crust eroded during the rift's activity (Fig. 2.11). y is given by the equations:

[2] $2.2y + 3.3z = x \times 2.8$
[3] $y + z = x$

which, as in the preceding case, respect the isostatic equilibrium between an initial state (a) and a final state (d) (Density of the basement of the crust: 2.8 g/cm³; density of the sediments 2.2 g/cm³; the density chosen for the mantle (3.3 g/cm³) assumes that the mantle has resumed its normal thermal state).

If $x = 13.5$ km (a maximum value which cannot be reached), then y (the thickness of the sedimentary layer) is 6 km. If $x = 5$ km, which is more likely, then $y = 2.25$ km.

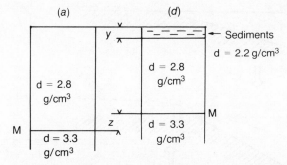

Fig. 2.11 The effects of contraction of a continental rift on its crests and flanks.

This reasoning and this calculation are valid not only for the continental shelves on the flanks of the old continental rifts but also for some intracratonic basins where rifts have aborted.

• In the first case (the continental shelf), the sedimentary layer forms a prism thickening seawards, i.e. where crustal erosion has been greatest (former rift crests).

• In the second case (the intracratonic basin), the cover thickens towards the centre of the basin. The graben belonging to this rift and filled with older sediments ought also to be found buried beneath the deposits accumulated during the contraction of the rift.

Consequently the erosion of the continental rifts and the thermal contraction of the continental lithosphere can only explain a total subsidence of some 2 or 3 km. Other geodynamic phenomena must therefore come into play on the continental margins, where the sediments are much thicker.

4 The effects of the sedimentary load

The sedimentary load is not a primary cause of subsidence, but it contributes to it. When sea-water (density 1.05 g/cm³) is replaced by sediments (density 2.2 g/cm³) a regional isostatic readjustment and a lowering of the crustal basement must take place. If a sedimentary layer 1 km thick were to be taken from a continental shelf or from a continental rise, the Moho would rise by about 0.5 km (the removed sediments being partly replaced by 500 m of water).

It will be recalled (Ch. 1, Fig. 1.7) that the application of weight to the lithosphere results in the formation of a depression surrounded by a broad zone of arching. This arching is accentuated when the load increases, i.e. when the deposits thicken in the case of a sedimentary basin. The fringes of the basin are thus uplifted whilst its centre is lowered. The consequences of such changes are very small in the oceanic environment. On the continental shelf, however, where the waters are always shallow, this uplift causes the emergence and erosion of the beds previously deposited on the edges of the basin. This explains the arrangement of the rock outcrops on many continental shelves and in the adjacent coastal basins (Fig. 2.12). The stratigraphic sequence towards the continental slope is fairly complete up to the present. In contrast, towards the coast and on the coastal plains beyond (which now are the exposed continuation of the continental shelf), older and older rocks are found which have been truncated by erosion. Paradoxically in this case, marine regression is an effect of subsidence.

Similar features occur on the oceanic side beneath the continental rise. Here, however, the transition between the ocean and the continent (which is itself a fossil

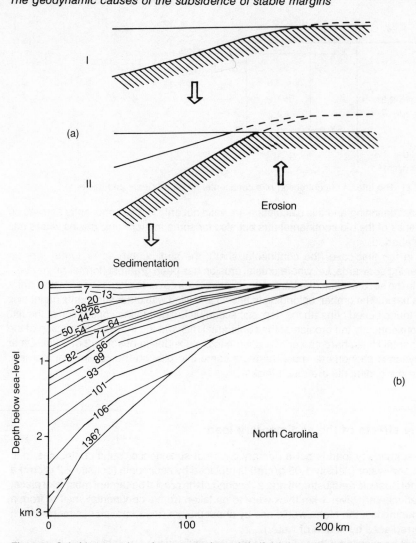

Fig. 2.12 Subsidence and erosion on a continental shelf. (a) theoretical diagram; (b) example of the continental shelf east of North Carolina, USA. (The figures indicate the absolute ages of the layers.) (Partly after Sleep and Snell, 1976.)

plate boundary) is also a mechanically fragile zone. It is occupied by a bunch of normal faults along which movement occurs throughout the sedimentation. Thus the subsidence of the continental rise occurs to a certain degree independently of that of the shelf. But this additional complication does not alter the direction of the vertical movements determined by the weight of the sediments. It merely means that the size of the movement on the continental side has to be calculated separately from that on the oceanic side.

5 The possible effects of metamorphism of the deep crust

The heat flow increase observed in the continental rifts is necessarily linked to an increased temperature in the crust and especially at its base. This may cause thermal metamorphism which in turn causes an increase in the density of the metamorphosed rocks and a consequent subsidence through isostatic reaction.

This hypothesis however encounters a serious difficulty. All the geological data make it quite clear that the base of the continental crust *already* has a high grade of metamorphism, well before rifting starts. An increase in temperature comparable to that noted on the continental rifts cannot therefore have important metamorphic effects. It may happen that a number of local or regional petrological modifications occur at depth and add their influence to other processes of subsidence. But this phenomenon on its own cannot account for the subsidence of the stable margins.

6 The thinning of the continental crust beneath the stable margins and the continent–ocean transition.

Seismic refraction shows that the Moho rises fairly gradually from the continental land masses (with their thick crust) up to the basins (with thin oceanic crust) that border the stable margins (*cf*. (c) p. 9). The lowering of the basement on these margins can therefore be interpreted as an isostatic consequence of the crustal thinning. One of the main causes of subsidence should therefore be sought amongst the factors responsible for the formation of an 'intermediate crust' in the continent–ocean transition zone.

(a) The problem of the magnetic quiet zone

The nature of this 'crust' is not known for certain and, at most, hypotheses may be formulated which will be verified or contradicted one day by oceanic drilling data.

Throughout the world's oceans the oceanic crust is marked by linear magnetic anomalies (amplitude 200–600γ), which are also isochrons (Ch. 1). Nearing the stable continental margins, however, this anomaly field often gives way to a much less disturbed zone which is some 100–400 km wide. This is called the 'calm anomaly zone' or '*magnetic quiet zone*' (M.Q.Z) where the size of the magnetic anomalies, where they exist at all, does not exceed 40–100γ on average (Fig. 2.13).

This M.Q.Z. has greatly intrigued geophysicists. It was first brought to light in the north Atlantic, east of the United States, and then in the south Atlantic. It has also been observed in the Norwegian Sea, which opened up only 60 m.y. ago and even in the Red Sea, where the first oceanic crust is less than 5 m.y. old. The first explanation proposed was that the M.Q.Z. is situated on ocean crust formed during a geological period when the direction of the earth's magnetic field stayed constant. This explanation has no general applicability because the M.Q.Z. is a different age in different areas. In the same way, their extensive global distribution demonstrates that these zones cannot all have been formed by oceanic accretion near the equator where the effects of thermoremanent magnetism in the crustal rocks are very weak.

More general interpretations have therefore been sought. For some geophysicists the M.Q.Z. represents a sector of ocean crust which underwent thermal metamorphism after its formation which could have removed its magnetic 'imprint'. But how could this explain that the phenomenon occurs only along a narrow oceanic band (lying close to the continent) which is bordered by strongly-marked anomalies that are undoubtedly isochrons? Others have suggested that evidence of slow oceanic opening can be seen in the M.Q.Z. where the magnetic imprint of the lithospheric accretion is blurred and 'averaged out'. Lastly the M.Q.Z. has also been interpreted as an area of very thin

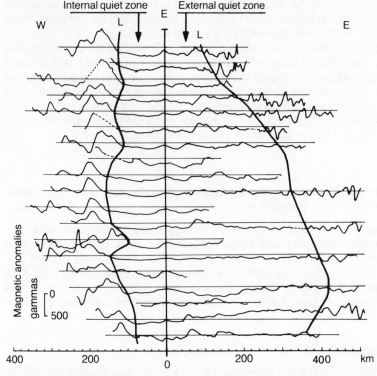

Fig. 2.13 The magnetic quiet zone off the New York area. The different profiles have been aligned according to the position of anomaly E. The major anomalies observed on the continental side may be explained by the emplacement of magmatic intrusions during rifting. (After Rabinowitz, 1974.)

continental crust. In this case it would form the deepest part of the stable continental margin where the substratum of the sediments is not oceanic in nature.

In fact, the M.Q.Z. has two sectors – at least in the north-west Atlantic where it has been most closely studied. On the continental side the anomalies are very weak. On the other side, in contrast, clear anomalies may be observed which form a kind of transition towards the disturbed field in the 'true' oceanic zone. These two parts are separated by a special anomaly, of about 100γ which is called the E anomaly (Fig. 2.13). Thus the various hypotheses proposed to account for the M.Q.Z. may be reconciled if the sector nearest the continental margin is considered to have a basement of a continental nature, whilst the other sector is probably oceanic in character.

This interpretation implies that the continental crust gradually thins out towards the ocean and the cause of this must therefore be found. As stated already (*cf*. §3, p. 27), the subaerial erosion of the lips and flanks of the continental rift can, in the end, remove only about 5 km of crust. Other hypotheses have therefore been proposed to account for this crustal thinning.

(*b*) The crustal creep hypothesis (Fig. 2.14)

Figure 2.14(a) represents a theoretical situation which occurs when two plates diverge and the first new oceanic crust appears between them. This puts two lithospheres of different thickness into contact. It also puts into contact a continental crust (on the left) and the oceanic mantle (on the right) between the levels M and M' (representing the oceanic Moho and the continental Moho respectively).

• The surface SC, which includes the continental Moho M′, is a surface of equal pressure wholly within the mantle. Above this surface the vertical distribution of densities differs in the continental milieu from that in the oceanic milieu. Thus, on any given horizontal surface between SC and the surface of the crust, pressures differ on either side of the 'wall' along which the continental and oceanic lithospheres come into contact. Above SC, the pressure decreases much more rapidly in the oceanic milieu (mantle density 3.3 g/cm³) than in the continental milieu (crustal density 2.8 g/cm³). In other words, the pressure Pc always exceeds the pressure Po at any given depth, and the difference can amount to 800 bars.

• During the beginning of the evolution of a stable continental margin (at the 'continental rift' stage) and at the very start of oceanic accretion, the temperature of the lithosphere rises markedly. Its mechanical properties may be modified as a result. Some geophysicists have suggested that the lower part of the continental crust could 'creep' under these conditions from the high pressure to the lower pressure zones, i.e. towards the oceanic lithosphere that is being formed. This hypothetical phenomenon is called 'hot creep'. If this displacement of crustal material really does occur, it should prompt a local rise in the Moho beneath the young continental margin. It should correspondingly bring about a marked regional subsidence leading to the formation of fault-troughs (Fig. 2.14(b))

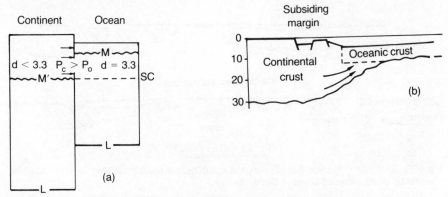

Fig. 2.14 Hot creep on a stable continental margin. M and M′: Moho; d: density expressed in g/cm³; SC: surface of equal pressure. (After Bott, 1971, simplified.)

This is an attractive hypothesis in so far as it explains at once the subsidence of the stable margins, the gradual thinning of the continental crust towards the ocean and certain similarities between the 'intermediate' crust and the deep zone of the continental crust (Ch. 1). Unfortunately, the model still remains very conjectural at present because no observations or geophysical measurements made hitherto can test its validity.

(c) Thinning by extension (Figs. 2.15 and 2.16)

The fault-trough on the continental rift marks the site of active early subsidence (*cf.* §1, p. 19). There, the continental crust becomes several kilometres thinner through the effects of crustal extension that is shown on the surface by synsedimentary normal faulting.

The evidence suggests that this feature becomes more extensive when the two lips of the rift begin to diverge and oceanic accretion starts. What then happens can be envisaged by magnifying the movements observed in the continental rift and by studying what is happening at present in the Red Sea (Figs. 2.15 and 2.16). The gradual rise of the Moho from the continent up to the true oceanic crust and the subsidence of the deep margin, which both result from crustal thinning caused by crustal extension,

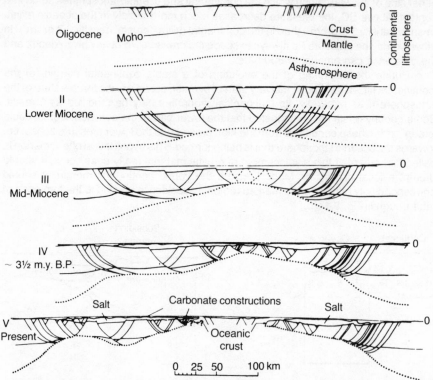

Fig. 2.15 Evolution of the Red Sea showing the role of normal faults in causing the thinning of the continental crust. (After Lowell and Genik, 1972.)

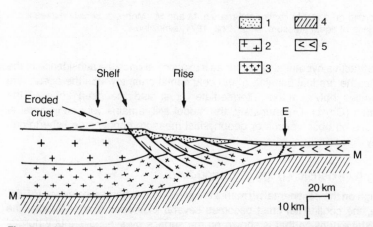

Fig. 2.16 Model of the thinning of the continental crust by extension on the stable continental margins. 1: sediment; 2: upper continental crust; 3: lower continental crust (probable granulite facies); 4: mantle; 5: oceanic crust; M: Moho.

can be explained by the gradual tilting of enormous crustal blocks along concave normal faults.

This 'model' would also explain a number of the characteristics of the 'intermediate crust':

• Its similarity to the oceanic crust. Extension facilitates the intrusion of basalt magmas from the upper mantle which often give rise to magnetic anomalies (Fig. 2.13).

• Its affinities with the deep continental crust (Ch. 1). The rotation of the crustal blocks brings the basal layers of the crust closer to the surface (Fig. 2.16). These are probably composed of high-grade metamorphic rocks (especially with granulite facies) (Ch. 1). Conversely, the granulitic basement rocks observed in fold mountains and often associated with flyschs (deep turbidites) may be interpreted as remnants of the intermediate crust of a former stable margin that has been deformed by plate collision. In this case these basements would have much the same significance as the ophiolites that are believed to represent the remains of the oceanic crust (Ch. 6).

Notes

1. Note that no margins older than the Trias are known, but that much older margins have been incorporated into the fold mountains.
2. The time constant is the time required so that only 1/2, 7/8th of the total subsidence remains to be accomplished. (This is about 1 km in practice.) The subsidence curve approaches a straight line if the depth is expressed as a function of the square root of the age of the lithosphere.

Further reading

Beaumont C., 1978. 'The evolution of sedimentary basins on a viscoelastic lithosphere: theory and examples' *Geophys. J. R. Astr. Soc.*, **55**, pp. 471–497.
Bott M. H. P. (ed), 1976. 'Sedimentary basins of continental margins and craton', *Development in Geotectonics*, **12**. Elsevier Publishing Company, Amsterdam, 314 pp.
Illies J. H. and **Fuchs K.** (eds), 1974. 'Approaches to Taphrogenesis', *Proceedings of an Internat. Rift Symp.*, Karlsruhe 13–15 April, 1972. E. Schweizerbart'sche Verlagsbuchhandlung, Stuttgart, 460 pp.
McKenzie D., 1978. 'Some remarks on the development of sedimentary basins', *Earth and Planet Sc. Letters*, **40**, pp. 25–32.
Parsons B. and **Sclater J. C.**, 1977. 'An analysis of the variation of ocean, bathymetry and heat flow with age', *Jour. of Geophysic. Res.*, **82**, pp. 803–27.
Ramberg I. B. and **Neumann E. R.** (eds), 1978. *Tectonics and Geophysics of Continental Rifts*. Reidel Publishing Company, Dordrecht, 443 pp.

Chapter 3
The structural and sedimentary evolution of stable margins

The first part of this chapter proposes a model which synthesises the structural and sedimentary evolution of the stable margins resulting from the gradual opening of an ocean. The chief stages in this evolution are marked by the 'continental rift', 'Red Sea', 'narrow ocean' and 'Atlantic' phases (§1). Then the model is completed by a brief description of the transverse structures on these margins and by the definition of a second family of stable margins caused by the strike–slip motion of two plates along a transform fault (§2). Finally, the last part of the chapter gives a more detailed description of the geological and geomorphological evolution of the continental shelf (§3), the continental slope and submarine canyons (§4), and of the continental rise (§5), as well as of the ways in which a stable margin may be transformed into an active margin (§6).

1 Model of the evolution of stable margins created by the divergence of two plates

The synthesis presented in summary in this paragraph is based essentially on the results of deep ocean drilling in the D.S.D.P. and the I.P.O.D. programme.[1] During the last decade some thirty drills have been established on the stable margins or in their immediate vicinity. The sedimentary evolution of these margins can be broadly reconstructed from the analysis of the core specimens. These drills have, however, only rarely reached the bases of the sedimentary sequence and so the start of the evolution can only be reconstructed from the study of the present-day continental rifts.

The four stages which can be broadly distinguished in the evolution of the margins created by lithospheric divergence, correspond to successive phases of ocean opening. These are called the 'continental rift' stage (already described in part in the previous chapter, §1), the 'Red Sea' stage, the 'narrow ocean' or internal sea stage and the 'Atlantic' stage.

(a) The 'continental rift' stage (Fig. 3.1(a): see also §1, p. 19).

At this stage the debris eroded from the land masses is mainly transported out of the fault-trough because of the form of the rift and its flanks. Sediments with *continental facies* (widespread fluvial or lacustrine deposits), *lagoonal facies* (evaporites) or *shallow marine facies*, are laid down in the central graben or half-graben. Sedimentation takes place in a restricted environment and organic matter is thus preserved, which in some cases is accompanied by *sapropelic* and petroligenous

Model of the evolution of stable margins created by the divergence of two plates

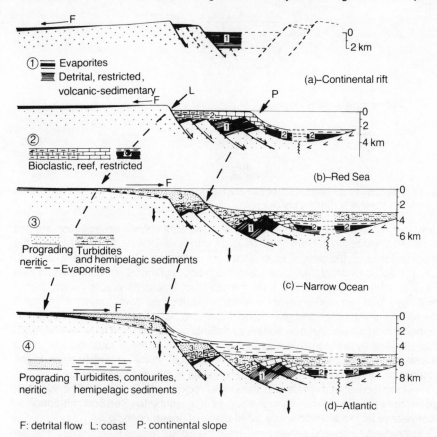

F: detrital flow   L: coast   P: continental slope

Fig. 3.1 The four principal stages in the evolution of a stable continental margin (the case of the European Atlantic margins).

formations. Lastly, volcanic activity occurs during the sedimentary sequence and takes the form of intercalated *lava-flows* and *volcano—sedimentary beds*. The sequence accumulated within the graben may reach a total thickness of several kilometres (3 km in the Rhine Rift Valley and 5 km in Lake Baikal in Siberia).

The fault-trough is often divided into two as in East Africa or in the Massif Central in France. In this case, if the rift changes into a zone of oceanic opening, one of the two grabens is 'aborted' whilst the other becomes the boundary between two divergent plates. The continental margins formed after such an evolution are thus different on either side of the new ocean. The simplest side has a half-graben structure, whilst the more complex side also contains the aborted graben buried beneath marine sediments (Fig. 3.2).

(b) The 'Red Sea' stage (Fig. 3.1(b))

Plate divergence has already begun at this stage and a narrow zone of oceanic crust has been created between the two margins that are being formed. The sea has already invaded the axis of the basin but communications with the open ocean are still difficult because the basin remains long and narrow.

• On the continental side, the lips and flanks of the original rift are preserved in such

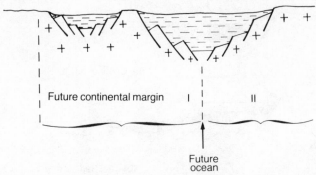

Fig. 3.2 Diagram of a continental rift divided into two, and the destiny of the system in the case of an evolution towards a stable margin.

a way that the products of subaerial erosion and atmospheric precipitation are still evacuated towards the outside of the basin as in the previous stage. (The Nile, not the Red Sea, takes up the detrital sediments from north-east Africa, for example.)

• An initial continental shelf grows up on both edges of the basin. In the present Red Sea, this shelf basically results from the action of limestone-producing organisms which have found an environment there that facilitates their development. Coral reefs grow up on the shelf edge and give rise to *reef limestones*. These form a kind of dam in the shelter of which bioclastic carbonate sediments are deposited. Thus the continental slope grows up by reef construction which compensates for the gradual lowering of the substratum. This slope can, however, undergo local slumping which leads to the formation of calcareous turbidites that accumulate in the deepest zones.

The formation of both reefs and carbonate rocks on the young continental shelf presupposes a warm climate. Though it occurs at present in the Red Sea, this condition is not necessarily linked with every initial stage of ocean opening.

• In the deep zones (more than 1 km deep) the waters remain stagnant because the basin is narrow and communications with the open ocean are difficult. It is a restricted, reducing, sedimentary environment which preserves the organic matter that falls from the upper layer of water where biological activity is intense. *Sapropels* and *Black shales* are formed in this way.

• Sea-water evaporation is not compensated for by the arrival of fresh water and its salt content can thus increase. Dense brines may sometimes form which then stagnate at the bottom of the basin. It is unlikely that these brines can precipitate directly. But, if communication with the ocean is broken for a time, the isolated arm of the sea may evaporate completely and form *evaporites*.

• Lastly, the vulcanicity seen on the borders of the marine basin and more especially the formation of oceanic crust in the basin axis are both linked to considerable hydrothermal activity. This increases the metallic salt content of the sediments which is mainly precipitated as sulphides. Metal deposits of great economic importance may thus be formed.

• The 'Red Sea' stage is also characterised by great subsidence. The contraction of the continental lithosphere has, as yet, hardly begun because of the proximity of the zone of oceanic accretion. But the extension of the continental crust, which began in the preceding stage, increases considerably. The crustal blocks rotate along curved faults which tend to become nearly horizontal lower down (*cf.* (c) p. 33 and Fig. 2.15). The movement of these blocks is compensated for on the surface by coral-reef building (Fig. 3.1(b)) or gives rise to half-grabens which are immediately filled with sediments (Fig. 3.3).

Model of the evolution of stable margins created by the divergence of two plates

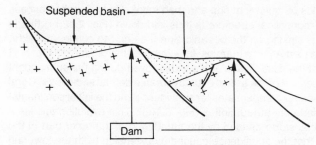

Fig. 3.3 Dams and suspended sedimentary basins brought about by the rotation of crustal blocks during the evolution of the margin.

• This stage of stable margin evolution is essential in the *formation of hydrocarbons*. Sapropels and black shales rich in organic matter are petroligenous rocks, whilst the calcareous turbidites intercalated in the deep sediments, and the carbonates on the continental shelf, form good reservoirs. The subsequent subsidence of the margin allows the burial of the sediments which are then subject to the pressures and temperatures needed to change the organic matter they contain into oil or gas. Lastly, as a result of the weight of the sediments, the evaporites are mobilised throughout the development of the margin. They give rise to diapir structures which always greatly facilitate the formation of hydrocarbon deposits.

(c) The 'narrow ocean' stage (or 'internal sea' stage) (Figs. 3.1(c) and 3.4).

The obliteration of the morphological barrier formed by the lips of the continental rift is the event which divides this stage from its predecessor. This results from the contraction of the margin which cools and lowers as it gets further from the zone of oceanic accretion. The sea floods over the flanks of the old rift which were still dry land at the 'Red Sea' stage.

• Detrital sediments are no longer directed away from the marine basin, chiefly because the bordering hills have disappeared. The deep zones are thus gradually buried by *turbidites* which are intercalated with *hemipelagic sediments*. The sea, however, is still too narrow for a general oceanic circulation to become established and the sedimentary environment thus remains relatively restricted. Particles of organic origin that are deposited are still therefore preserved. But they are diluted by the detrital input and thus the petroligenous potential of these sediments is generally lower than that of the deposits laid down in the 'Red Sea' stage. It sometimes happens, however, that the proportion of preserved organic matter increases considerably when the detrital influx itself from the continent contains abundant debris of vegetal origin. Then more 'black shales' are formed once again (which in this case are really 'black turbidites'), but their origin and depositional conditions differ from those laid down previously.

• The submergence of the flanks of the rift at the very beginning of the 'narrow ocean' stage may be accompanied by the deposition of a *shelf evaporite sequence* at the base of the overstepping beds. These should be clearly distinguished from beds of similar nature in the previous stages. The older beds are preserved in deep basins on the stable margins whilst the others, generally formed later, are found at the base of the cover on the continental shelves.

• After any such evaporite deposition, the continental shelf is formed chiefly by 'progradation' (*cf*. (*b*) p. 44). As the edge of the continent is lowered by erosion during the continental rift and Red Sea stages, neritic sediments build up a wedge which thickens seawards. It builds up quickly at first and then more slowly as subsidence is reduced. A cover with a monoclinal structure is thus laid down (Fig. 2.12(b)) that is dictated by greater crustal sinking on the oceanic side than on the continental side.

• In many cases (in the east Atlantic margins, for example), the organisms that built up the continental shelf at the 'Red Sea' stage are destroyed during the subsequent stage. This is probably because of the arrival of unfavourable detrital sediments. Coral reef construction thus no longer compensates for subsidence and the initial continental shelf is lowered and buried beneath turbidites (Fig. 3.1(c)). This evolution implies a *recession of the continental slope* caused by the foundering of the outer part of the shelf. It seems therefore that the subsidence caused by extension (with the lowering and rotation of crustal blocks) has not entirely ceased at this stage in the evolution of the stable margins.

• In other cases (in the Atlantic margins of the United States, for example), coral reefs are still built up during the 'narrow ocean' stage (Fig. 3.4; see also Fig. 2.1). A kind of dam is thus formed and enormous thicknesses of sediments are accumulated behind it.[2]

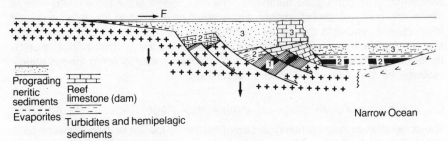

Fig. 3.4 The 'narrow ocean stage' in the case where a coral barrier is built upon the shelf-edge (the case in the Atlantic margin of the United States).

• The continental slope of a narrow ocean has an important and varied sedimentary role. In different areas and at different depths it can be:

(i) A *transit zone*. Turbidity currents follow submarine canyons before depositing their load on the continental rise or in the deep ocean basins (*cf*. (*a*) p. 48).

(ii) A *'progradation front'* when the sedimentary slope bordering the continental shelf grows quickly out to sea (*cf*. (*b*) p. 44).

(iii) A zone of very slow pelagic sedimentation which is protected from detrital input. In this case, the condensed sequences that are familiar to geologists are deposited. Here a few metres of sediments sometimes contain microfossils from several geological stages.

(iv) Lastly, *an erosional zone* where subsidence so increases the angle of the floor that it becomes too steep for the sediments to remain stable upon it.

From the 'narrow ocean' stage onwards, the weight of the sea-water (as it deepens) and of the sediments can bring about the mobilisation of the evaporites deposited previously and, with it, the appearance of the first diapir deformations affecting the deep margin.

(*d*) The 'Atlantic' stage (Fig. 3.1(d))

When it is fully developed, the ocean extends over a sufficiently wide area for climatic contrasts to develop from one side to another. Access to cold water sources (in the areas near the poles) sets up an oceanic circulation that completely changes sedimentation conditions. The deep waters are renewed by the sinking of cold, dense and well-oxygenated water and organic matter is most often destroyed by oxidation. Moreover, deep currents can take up sedimentary particles and transport them great distances. Some areas are thus subject to erosion (and previously deposited sediments are reworked), or are deprived of sedimentary input for long periods. In contrast, the sedimentation rate is greatly increased where the currents slow down and drop their load. Accumulation occurs especially on the continental rises (Fig. 3.1(d)). They not only receive *turbidites*, but also the bulk of the solid load of the deep currents (which are also called 'contour currents'). Thus the finely laminated sediments called '*contourites*' are formed.[3]

Thermal subsidence is considerably slowed down at this final stage in evolution of the margin:

• The *continental shelf*, with a basement which is hardly sinking any more, is henceforth subject to the effects of marine transgressions and regressions (with alternating periods of subaerial erosion and marine sedimentation) which are eustatic in origin. They are dictated either by the climatic changes at the close of the Cainozoic (the great glaciations) or by variations in the rate of oceanic expansion. (Rapid accretion on the axes of the mid-ocean ridges gives rise to a wide swelling on the ocean floors and causes the sea to overflow on to the edges of the continents). The amount of progradation then depends on the volume of sediments brought in (*cf.* §3, p. 43). The *continental slope* on the 'nourished' margins can then become a sedimentation slope whilst the continental slope on a 'starved' margin remains a zone of condensed sedimentation and erosion.

• The *growth of the continental rise* greatly depends on its situation within the ocean. For example, the American Atlantic margin receives more contourites than the European margin because it is more exposed to the action of deep currents.

• *Diapirism*, lastly, which began during the previous stage, continues through the Atlantic stage because of the weight of the sediments accumulating on the continental rise.[4]

2 Strike–slip margins and transverse structures on the stable margins

The evolution described in the previous paragraphs is the result of the gradual separation of two plates. But there is a further category of margin that is also stable, which results from strike-slip movement along a transform fault putting continental crust into contact with ocean crust (Fig. 3.5, margin 2). This second 'family' of stable margins is less frequent than the first which is created by lithospheric divergence (Fig. 3.5, margin 1). The clearest example of this second family is provided by the northern gulf of Guinea.

The passage from one type of stable margin to another along a continental edge explains a number of transverse structures that are seen in the sedimentary basins on these margins.

(*a*) The subsidence on the margin is greatest on the graben of the old continental rift and decreases towards the dry land (Ch. 2). The geometry of the initial fracture pattern which preceded the drifting of the two continental blocks and the birth of the new ocean between them, thus closely governs the distribution of the subsidence zones on the margins of this ocean. For example (Fig. 3.5), when a rift is crossed by a transform fault, it may be foreseen that the subsidence zone corresponding to it on the future margin will be offset in approximately the same way.

(*b*) The outline of the initial fracture that is marked by the continental rifts and intra continental transform faults does not occur by chance. When tensional stresses act on

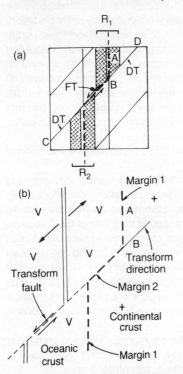

Fig. 3.5 The idea of transform direction and strike–slip margins (explanations in the text). D.T.: transform direction; F.T.: transform fault; R_1–R_2: initial continental rift; margin 1: stable margin created by lithospheric divergence; margin 2: stable margin created by a strike–slip movement along a transform fault.

the plate, the rifts seem to be installed preferably on the weak zones of the crust, i.e. directly upon previous geological structures. In Fig. 3.5(a), a continental plate is assumed to be crossed by an old network of fractures trending N-S and NE-SW. As a result of extension, rifts are established on the N-S faults whilst the old fracture CD is reactivated and acts as a transform fault FT between the two ends of the rifts R_1 and R_2. (This situation occurs at present in France with the Rhine rift system, the Jura, the Bresse trough and the Limagnes.)

(c) Right and left of the segment R_1–R_2, the fracture CD is not really a transform fault. But it is the continuation of such a dislocation and is designated 'transform direction', DT, for that reason.[5] It will be noted that the transform direction does not take part in the sliding motion, but may, on the other hand, act as a system of normal faults. During the activity of the continental rift, it separates a rising crustal compartment, A, from another compartment, B, which remains stable (Fig. 3.5(a)). In contrast, when the rift contracts (Fig. 3.5(b), the same fault separates the stable compartment, B, from an area, A, which is subsiding.

This explains how the structural and sedimentary evolution of the stable margins is determined not only by normal faults parallel to the ocean–continent boundary (i.e. parallel to the former axis of the rift), but also by transverse dislocations which offset subsidence zones and delimit some sedimentary basins.

3 The continental shelf on stable margins and its evolution

The upper limit of the present continental shelf is the coastline. But this border merely has a temporary geological significance and many recently-emerged plains and coastal basins belong, by their structures and sedimentary histories, to the continental shelf alongside them.

In contrast, the lower limit of the shelf has the value of a geomorphological and structural boundary. It forms the shelf-break, a more or less pronounced break of slope which is most often situated at about -180 m (the 100 fathom limit), but is sometimes shallower (-120 to -130 m in the Mediterranean) and sometimes deeper (up to -500 m).

The shelf varies greatly in width. It is narrow or very narrow along mountainous coasts (in the French Maritime Alps, for instance) but it widens considerably in front of low coasts (and sometimes reaches several hundreds of kilometres). The world average is 70–80 km. These differences may be explained by the evolution of the margins as described in §1, p. 36. Narrow shelves bordering areas of rugged relief generally belong either to youthful margins ('Red Sea' stage) or to rejuvenated margins (that have undergone recent tectonic deformation). On the other hand, the wide shelves forming the submarine continuations of extensive coastal plains are evidence of a much more advanced stage of evolution (the 'narrow ocean' or 'Atlantic' stage).

(a) Neritic sediments

• The continental shelf is a place where life abounds. The shallowness of the water allows the light to penetrate which chlorophyllous organisms need. The continent nearby is the main source of the nitrates and phosphates that are essential to biological activity. Lastly, the constant movement of the waters provides permanent oxygenation of the environment. The continental shelf is thus a favoured site for the accumulation of *bioclastic* (or biogenic or organogenic) sediments. The *benthos* provides the most important contribution to this sedimentation of biological origin. Algae, and creatures living in symbiosis with algae such as reef corals, abound in the shallowest areas (from 0 to -50 m or to -100 m according to the transparency of the water). Molluscs, Echinoderms, Bryozoa, Brachiopods, etc. are found at all depths as far as the top of the continental slope. In warm and temperate climates this activity ends in the production of *carbonate* sediments (mainly limestones). In cold climates the insoluble carbonate – soluble bicarbonate chemical balance is displaced. Limestone formation is impeded and the biogenic sedimentation becomes more *siliceous* in character (with diatomite formation, for example).

• The *residual* sedimentary particles have quite a different origin. They are formed by *submarine erosion*, i.e. the reworking of rocks (including unconsolidated sediments, of course) that were formed in different conditions from those prevailing at present. The rocks eroded in this way may be old (as, for instance, when a limestone exposed on the shelf is gradually destroyed) or they may be more recent. (An unconsolidated formation laid down during a Quaternary regression is easily modified, for example. Most of the north Atlantic continental shelves are covered with such residual deposits and prove the occurrence of low sea-levels and cold climates during the Pleistocene.)

The most typical residual sediments are precisely those produced by the modification of continental formations at the beginning of the marine transgressions. Alluvium, periglacial deposits, moraines or wind-blown dunes, which were laid down when the continental shelf was dry land (through a eustatic lowering of sea-level) are taken up again to varying degrees by the sea when it re-occupies its former territory.

They then become the classic 'basal conglomerates' by which almost every neritic sedimentary cycle begins.

• *Detrital* sediments, on the other hand, are produced by subaerial action: coastal, eolian and fluvial erosion and accumulation, etc. Thus, for example, pelites (that are more or less calcareous in nature and later become marls), or siliceous sands, may be deposited according to the hydrodynamic energy within the environment.

• Lastly, *authigenic* sediments (phosphorites, glauconites, etc.) are freshly formed on the floor wholly or partly from substances dissolved in sea-water.

These varied sediments are generally very mobile whatever their origin. Considerable hydrodynamic energy is located on the continental shelf. In different cases, tidal currents, swell currents and internal waves (on the shelf break) find a favoured environment for their activity. Thus the neritic sediments are generally *sandy* and more rarely *pelitic*. If, as in the case on margins that have reached maturity, the shelf is not subsiding, these sediments form a cover that is being gradually displaced towards the continental slope.

(b) Explanatory model of 'progradation' (Fig. 3.6)

The term 'progradation' expresses the idea of a seaward extension of the continental shelf by successive sedimentary deposition on its front. In the model proposed in this paragraph, the phenomenon is governed by five factors:

1. The depth (H) of particle stability (LS)

For any given granulometric class (fine sands, for example), this depth will vary according to the seas and even to local conditions. The depth depends on the size of the currents and the swell, and hence on the form of the coast and the latitude. In principle, H varies with the depth of the shelf-break. If they are not so deep, the particles stay mobile and can be displaced oceanwards in the direction of the steepest slope. Conversely, where the water is deeper than H, these same particles become stabilised and form a sedimentary talus on the upper parts of the continental slope. For the sake of simplicity, H is assumed to be constant on Fig. 3.6.

2. The subsidence of the continental shelf

In the case where no subsidence occurs the very thin sedimentation occurs on the shelf itself but, on the other hand, it supplies the materials for rapid progradation. The deposition of beds S_1, S_2, and S_3 (Fig. 3.6(a)), pushes the shelf seawards. The granulometry G of the particles deposited decreases rapidly with depth.

In contrast, if the margin is subsiding, the depth H is kept constant by sedimentation on the shelf itself (Fig. 3.6(b)). Progradation is less rapid because a proportion of the sedimentary particles does not reach the shelf-break.

3. Absolute variations of sea level

According to the previous paragraphs, the decrease in water depth through a eustatic lowering of sea-level should bring about much erosion and the resultant re-establishment of the depth H (Fig. 3.6(c)). The sedimentary material that is taken up again in this way supplies the progradation front and gives rise to a thick bed R_1 which is formed, in part, from the previously deposited beds S_1, S_2 and S_3.

Conversely a rise in sea-level enables a thick bed to be accumulated on the shelf whilst only a relatively thin layer is deposited on the continental slope. Thus once again the depth H is re-established (Fig. 3.6(d)).

Naturally subsidence and eustatic factors can act together. For example, in Fig.

3.6(e)), it is assumed that the situation reached in Fig. 3.6(d) has been modified by a new eustatic regression (deposition of the R_2 beds), after a period of subsidence (deposition of the S_4 beds).

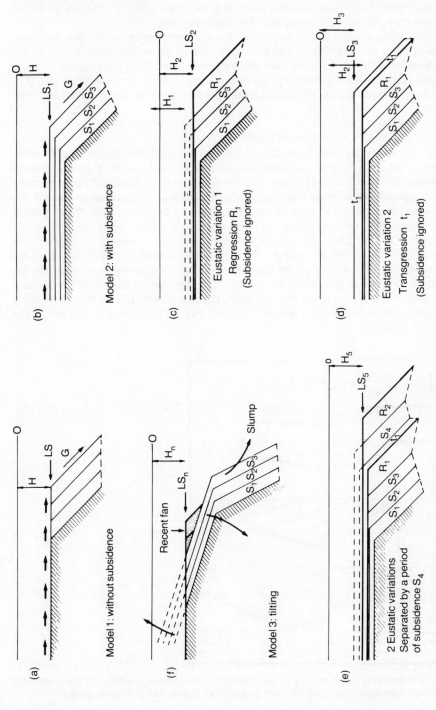

Fig. 3.6 Large-scale progradation on the continental shelves (explanations in text).

45

4. Tectonic movements

Figure 3.6(f) shows the effects of tilting. The uplifted part of the shelf (which is nearest the coast) has been eroded, whilst a new progradation wedge has been built up on the lowered part. At the same time the increase in the angle of the former talus S_1–S_3 can cause slumping which 'rejuvenates' the continental slope.

Such a movement can be the result of a local deformation of the margin. It will, however, be recalled (Ch. 2) that sinking caused by the weight of sediments on the outer edge of the shelf (on the oceanic side where the progradation wedge forms) is balanced by the uplift and emergence of its internal edge on the continental side (Fig. 2.12(a)). In other words, progradation itself contributes to the tilting described in Fig. 3.6(f).

5. The importance of sedimentary supply

The amount of progradation also depends on the quantity of sediments supplied to the continental shelf. Progradation is slow or ceases if no great quantity of detrital material crosses the shelf. This may be because the continent is not being eroded or because the materials are diverted down submarine canyons to the continental rise (*cf*. §4, p. 47). The continental slope is then only covered by a very thin film of pelagic sediments. In contrast, considerable progradation occurs on some 'nourished' margins. In this case the *continental shelf extends seawards throughout the geological history of the margin* (Fig. 3.7). Its external part can lie upon the *intermediate crust* or even on the oceanic crust. This, for example, is the case with the submarine delta of the Niger. As a result, it is important to distinguish large-scale progradation (Fig. 3.6) which only affects the upper continental slope, from small-scale progradation, which is much greater in volume (Fig. 3.7) and more closely akin to the growth of deltas.

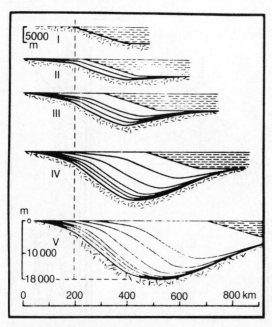

Fig. 3.7 Growth of a sedimentary accumulation by small-scale progradation on the continental shelf-edge (margin 'nourished' by major sedimentary inputs). (After Walcott, 1972.)

(c) The effects of emergence on continental shelves

When it becomes dry land, the continental shelf is subject to all the agents of subaerial erosion and deposition. When it is submerged again, it retains traces of the regression both in its sediments (residual deposits: *cf*. (*a*) p. 43) and in its morphology. The latter may include: *fluvial valleys* (filled with alluvium and recent deposits and also locally exhumed to form closed depressions); *karsts* (often well developed because the rocks with neritic facies on the shelf are most often carbonate in nature); *glacial troughs* in high latitudes; *marginal depressions* (excavated on the boundary between the basement and its cover which may sometimes exceed 100 or 200 m in depth); *platforms of marine abrasion* (developed at marine Quaternary stillstand levels); and *old relief features* (= rocky banks) (fossilised at their bases by modern deposits but eroded at their summits).

However, these relict features do not seriously alter the flatness of the shelf. Important topographic features are rare on this vast submarine expanse which slopes seawards at an average angle of only 0.07 °. These topographic features thus seem all the more remarkable, and they give rise to linguistic excesses: a depression a few tens of metres deep becomes a trench and the smallest mound becomes a bank.

The emergence of the shelf can have more lasting consequences by facilitating the *lithification of sediments*. In the event of a marine regression, the limestones previously deposited are subject to vertical circulation beneath the water table and are cemented by solution and carbonate precipitation – if they have not been eroded and scattered over the progradation front (*cf*. (*b*) p. 44). A kind of slab is thus formed (Fig. 3.8) which may lie obliquely across the stratification and covers sediments that are probably compacted but not cemented. Because of its resistance to erosional agents, the edge of the slab forms a cornice when the sea returns.

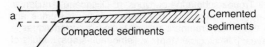

Fig. 3.8 Lithification of carbonate sediments on a continental shelf in the case of a eustatic fall in the sea-level a.

In conclusion, continental shelves may be divided into two groups:
(i) *Shelves built up by present-day sedimentation* (Fig. 3.6(b)). These usually belong to young subsiding margins or to margins rejuvenated by tectonic movements. (Submarine deltas are an exception, and so, in general, are the shelves 'nourished' by great sedimentary input.)
(ii) *Abrasion platforms*, which, in contrast, belong to old margins where subsidence has almost entirely ceased (Fig. 3.6(a)).

4 The continental slope on stable margins and its evolution

The continental slope is not as steeply inclined as is often believed. The world-wide average angle is 4° 17′. (It is 5° 20′ in the Pacific where active margins predominate; 3° 05′ in the Atlantic where almost all the slopes belong to well-developed stable margins; whilst it is 3° 34′ on the stable but still young margins of the western Mediterranean.)[6] Off the great submarine deltas (such as those of the Niger, the Mississippi and the Nile), the continental slope is a small scale progradation front (Fig. 3.7) and is even gentler (1° 20′ on average).[6] Its angle often decreases downwards and thus develops a broadly

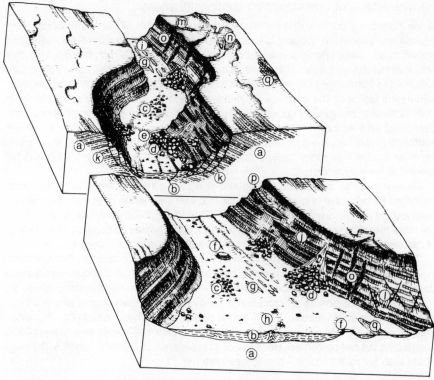

Fig. 3.9 The Les Stoechades and Saint Tropez submarine canyons. a: Oligocene–Aquitanian rocks; b: Pleistocene; c: blocks scattered or grouped on the canyon floor; d: slumps and block-masses at the foot of the slopes; e: threshold; f: remnant outlying buttes; g: current ridges; h: structure attributed to mud diapirs; i: caves; j: lithological steps; k: listric fault-steps; l: faults of tectonic origin; m: gullies, gorges; n: channels; o: ravines; p: crevasses caused by slumping; q: sparse outcrops beyond the canyon. (After the Groupe Estocade, 1977.)

concave profile. The continental slope and the continental rise generally merge about −3 000 m or −4 000 m or sometimes higher.

(a) Submarine canyons

In many areas of the globe the continental slope is dissected by submarine valleys. These are the famous 'canyons', whose nature and origin have long been one of the major problems of submarine geology and geomorphology. The word itself should not be allowed to be misleading: it is true that some canyons have steep walls, but they are most often only valleys with gentle slopes.

These submarine canyons along with the marginal shelves (*cf.* (*b*) p. 52) form the chief relief features on the continental slope and can be either submerged continental valleys or forms resulting from submarine erosion.

1. Submerged continental valleys

The continental shelf displays erosional forms that were excavated in the open air, were more or less filled in by alluvium and are now submerged (*cf.* (*c*) p. 47). But in certain cases the continental slope is itself also dissected by valleys and their morphology demonstrates beyond a shadow of doubt that they were formed by subaerial agents.

The western Mediterranean canyons provide the clearest examples here (Figs. 3.9 and 3.10).

In the 1950s, as soon as it was established that the submarine canyons in Corsica and Provence were former continental valleys, two conflicting hypotheses were formulated. Some geologists envisaged a great marine regression that exposed the whole of the continental slope to the atmosphere so that it could be eroded by torrential streams. Others, on the other hand, proposed that the gradual submergence of the continent and its drainage pattern could be explained by tectonic mechanisms.

First hypothesis: Continental flexure (Fig. 3.11)   The geology of Provence, the Maritime Alps and the adjacent submarine floors demonstrate that the western Mediterranean basin is of recent origin. At the close of the Eocene and probably even still in the lower Oligocene, the present sea was occupied by a 'Ligurian continent'. The basin which now separates Corsica from Provence only opened up in the upper Oligocene and Miocene. Such an evolution naturally implies a complete reversal in the direction of drainage and detrital transport. At first (in upper Eocene and lower Oligocene times), the direction was divergent in relation to the Ligurian continent, whilst later (from the end of the Oligocene to the present) it has converged upon the young marine basin.

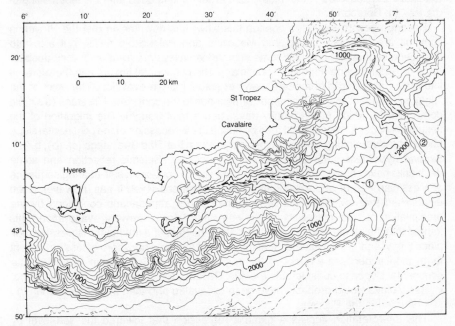

Fig. 3.10 Morphology of the continental slope off the Maures Massif. 1: Les Stoechades canyon; 2: Saint Tropez canyon. (After Gennesseaux and Vanney, 1979.)

This 'revolution' was a gradual process in reality. It was first interpreted by the 'continental flexure' hypothesis (Fig. 3.11). This involved a gradual uplift in the north (in the sub-alpine zones) which was accompanied by a lowering of the Provençal margin in the south. (A reversed but otherwise identical model went for the Corsican continent and its margin.) These two opposite movements imply the existence of a flexure which is assumed to be mobile but was always located on and merged with the coastline. Its migration brings about the gradual submergence of the coastline and its valleys. 'A Provençal canyon thus has three sections from its lower reaches to its upper reaches: a

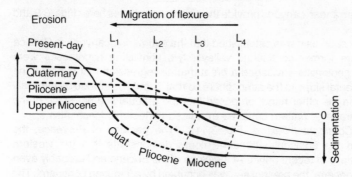

Fig. 3.11 Continental flexure and the gradual submergence of the submarine canyons. L_1, L_2, L_3, L_4: successive coasts, merged with successive positions of the continental flexure. (Diagram synthesised from the works of J. Bourcart, 1958.)

deep part excavated by Miocene fluvial erosion and flooded by the sea in the Pliocene; a middle part excavated by end-Pliocene erosion and flooded in the Quaternary; and a terminal part excavated in the middle Quaternary and flooded after the last melting of the ice.' (J. Bourcart.)

Recent research seems to confirm this view. It is now known that the Provence canyons are not only carved into Mesozoic and Palaeozoic rocks, but also into Oligocene sediments, and that some submarine valleys are filled up in their deepest parts by end-Miocene deposits (Messinian). The continental flexure can therefore be interpreted in the light of modern views of stable margin evolution. The uplift of the 'Ligurian continent' in the Oligocene corresponds to the continental rift stage (a simple dome initially, which subsequently developed a fault-trough). The migration of the flexure in Miocene times should then be linked to the recession of the continental slope and the gradual sinking of the crustal blocks during the 'Red Sea' stage (*cf*. (*b*), p. 37).

Second hypothesis: The end-Miocene regression   Seismic reflection and some deep drills have shown that the Mediterranean basin contains a salt-bearing sequence of Messinian age at about 4 000 m below present sea-level. It has, however, been established that this sequence was formed under atmospheric conditions by the precipitation of salts dissolved in sea water that later underwent evaporation. This implies that the Mediterranean dried up before the Pliocene marine transgression took place 5 m.y. ago. If it is assumed that the basin at that time had more or less its present form, then a necessary effect of the regression would have been to subject the continental slopes to considerable erosion. Because the slopes were relatively steep, incised valleys would probably have been excavated and then submerged when the sea returned in the Pliocene.

This interpretation implies a gigantic regression that lowered the sea-level by several kilometres, which was followed by a marine transgression of the same magnitude. But are such variations possible?

At present the Mediterranean has a volume of 4 240 000 km³. The net balance of sea-water entry and exit through the Straits of Gibraltar amounts to 1 500 km³ per year from the Atlantic into the Alboran Sea. This supply makes up for the evaporation from the Mediterranean. If it were to be cut off, then the basin would dry up in less than 3 000 years. This is almost instantaneous on the geological time-scale. In fact, the Straits of Gibraltar are shallow (500 m) and a slight tectonic movement there would be enough to sever communications with the Atlantic. The Messinian marine regression and the Pliocene 'inundation' can thus be explained in these terms. What is, however, in dispute in this interpretation of submarine canyons is the basic assumption: is it certain that the

Miocene Mediterranean basin was as deep as the present basin? In other words, were the slopes during the Messinian regression as steep as those observed today? This debate continues.

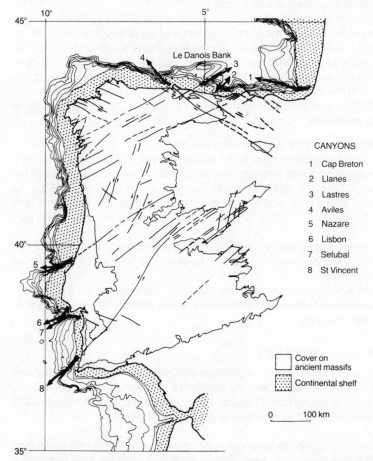

Fig. 3.12 Transcurrent and late-Hercynian faults in the Iberian Peninsula (after Parga, 1969) and the orientation of the chief submarine canyons (after Boillot *et al.*, 1974).

2. Valleys formed by submarine erosion

As regional studies continue, it becomes increasingly clear that almost all the submarine canyons are erosional forms situated on or near major tectonic dislocations. For example, the many and deep canyons seen on the Atlantic façade of the Iberian Peninsula (Fig. 3.12) mark the continuations of old faults and wrenches that have been recently re-activated. ('The gouf de Cap Breton' which seems to provide an exception, is, in fact, associated with the north Pyrenean frontal overthrust.)

• Under the sea, as in the atmosphere, erosion is facilitated in those areas where the rocks offer least resistance for tectonic or lithological reasons:
 (i) *direct structural control*: faults and, in a more general way, all tectonic dislocations cause local crushing of the rocks which are then more easily eroded there than elsewhere.
 (ii) *indirect structural control*: canyons can be excavated in recent unconsolidated

sediments or they can be formed on older rock outcrops which are easily eroded because of their petrographic nature. Any deformation of the sedimentary cover on the margin which causes the exposure of weak beds can thus prompt the development of a submarine valley upon them.

• Most of the great submarine canyons observed on the developed stable margins were excavated at the close of the main subsidence of these margins. They cannot therefore be explained by the continental flexure hypothesis, which on the contrary, implies appreciable sinking after their formation. Moreover, although it is quite conceivable that such a relatively narrow basin as the Mediterranean could dry out, it is quite implausible in the case of the major oceans which communicate easily with each other. It must therefore be acknowledged that the canyons on the old stable margins were formed by *submarine erosion*.

Many agents probably took part in their excavation:
 (i) *Turbidity currents* probably have some erosive power. They do, in actual fact, follow submarine canyons and often break submarine telephone cables in the valley-floors. The mere fact that these valleys are not themselves filled up demonstrates that turbid waters can rework unconsolidated sediments and 'maintain' the canyon, which is undoubtedly an important site of sedimentary transit.
 (ii) *Seismicity* is associated with 'active' tectonic dislocations which often guide the path of canyons as has been stated above. Earthquakes facilitate all processes of slumping and erosion;
(iii) Lastly, all kinds of other factors can make the canyon sideslopes recede if only the rocks exposed are sufficiently unconsolidated : sandy flows, mudflows, slow slides of sedimentary masses, etc.

(b) Dams and marginal plateaux

The continental slopes of the stable margins locally display breaks of slope and tabular surfaces which are transitional between the continental shelf and the continental rise or abyssal plains which they overlook. They occur at depths varying between -500 m and $-4\,000$ m and are called *'marginal plateaux'* (or 'marginal banks').

1. Some of these correspond to horsts or even 'micro-continents' which developed during rifting. The high structural zones, between the fault-troughs, then form major relief features. These are not necessarily removed by the subsequent subsidence of the margin but are merely lowered along with the sedimentary basins around them. They are mantled at the close of their evolution by neritic sediments (formed at the beginning of their submergence) which are overlain by pelagic or hemipelagic deposits (accumulated when the banks have sunk deeply).

As a result in this case the marginal plateau is created during the 'continental rift' stage.

2. The prolonged subsidence of the margin implies the permanent activity of normal faults, chiefly during the 'rift' and 'Red Sea' stages, but also during the 'narrow ocean' stage as well. The mobile blocks sink by rotation and give rise to structural relief features behind which sediments are accumulated (Fig. 3.3). The group formed by the pair composed of a 'suspended basin' and a 'dam' constitutes the second type of marginal plateau, which must be younger than the previous type.

3. The idea of the *'dam'* which has just been introduced can be applied more generally to every high structure which causes the development of a suspended sedimentary basin.

• Some dams are composed of *'barrier reefs'* which develop on the shelf edge and 'hold back' sediments between them and the coast (Fig. 3.4).

• Others are volcanic edifices which usually develop during the initial stages of margin evolution.

• Evaporites at the base of the sedimentary sequence on the margins often give rise to diapirs (*cf.* p. 41). These are sometimes arranged in 'salt walls' which form hills that can play the same sedimentary role as the volcanic chains in 'trapping' sediments on their continental side.

4. Lastly, some stable margins are deformed at some time during their history without becoming active margins in the real sense of the word. Old structures are reactivated in this case and marginal plateaux (horsts) can develop, which are entirely new from the geomorphological point of view (for example, the Galicia bank to the north-west of the Iberian Peninsula).

5 The continental rise on stable margins

The slope of the continental rise is both gentle and regular (from 1% to 0.1%). Its border with the abyssal plains is situated at an average depth of 5 000 m but without a well-marked break of slope. In front of the mouths of major rivers such as the Ganges and the Indus, for example (Fig. 3.13), the continental rise develops to such an abnormal degree that it buries most of the continental slope.[7]

Relief features are rare in this last and deepest morphological province on the stable margins:

(*a*) *A few sea-mounts* bear witness to volcanic activity which is extinct for the most part.

(*b*) *Deep submarine sedimentary fans* (or deep deltas) correspond to thickenings of the continental rise caused by a great accumulation of turbidites. They are crossed by a *'fan-shaped' valley* system (or a bird's-foot system) which extends from the canyons on the continental slope. They have the particular characteristic of branching as they go deeper. These are the divergent channels of the turbidity currents. Their cross-profiles show 'lateral levées' which were built up when the turbid waters spilled out from the main confining channel.

6 The transformation from a stable margin into an active margin

At the close of the evolution of a stable margin the lithosphere which was newly formed during the 'Red Sea' stage has cooled completely. The contrast between its relatively high density and that of the asthenosphere immediately below is enough for the 'aged' oceanic lithosphere to tend to sink into the asthenospheric mantle. In contrast, the lithosphere carrying continental crust is less dense than the asthenosphere on the whole and can therefore only remain on the surface of the globe.

In such conditions, decoupling could theoretically take place in the transition zone between the two kinds of lithosphere, i.e. on the ocean–continent boundary. According to some hypotheses, this is how subduction of the oceanic lithosphere beneath the continental lithosphere would begin. This would have the effect of changing the former stable margin into an active margin. Working on the basis that lithospheric density changes with time, geophysicists have calculated that this event ought to occur when the stable margin exceeds 180 to 200 million years in age. This would explain the

Fig. 3.13 The deep delta of the Indus in the northern Indian Ocean. The continental rise here takes on quite exceptional dimensions. Scale 1 cm = 95 km. (After Heezen and Tharp, 1964.)

absence of old oceans on the globe (i.e. more than 200 m.y. old), whereas rocks are known on the ancient continental 'shields' that were formed more than 3 000 million years ago.

The change from a stable margin into an active margin can, however, occur at any stage in the evolution of an ocean. The old plate boundary on the ocean–continent limit is a zone of weakness in the lithosphere, which is more or less mobile during the evolution of the sedimentary stable margin (Chs. 2 and 3). If compression occurs, this boundary can be re-activated and transformed into a plane along which the relatively light continental lithosphere is thrust over the oceanic lithosphere that is always denser. The subduction which starts off in this way will continue thereafter because of the density of the descending lithosphere which enables it to plunge into the asthenosphere under its own weight.

Whatever interpretation is adopted, a stable margin, whose evolution was determined by extension phenomena, can clearly be greatly altered by the effects of *compression* when the subduction of the bordering oceanic lithosphere starts. The normal faults of the stable margin then become reverse faults. The crustal blocks which were subsiding before can be changed into tectonic wedges and into overthrust nappes. This is the start of another story, that of an active continental margin (Chs. 4 and 5).

Notes

1. D.S.D.P.: Deep Sea Drilling Project, under the scientific auspices of the J.O.I.D.E.S. (Joint Oceanographic Institutions for Deep Earth Sampling). I.P.O.D. (International Phase of Oceanic Drilling) grew up when the J.O.I.D.E.S. was enlarged to include foreign partners of the United States.
2. The main hydrocarbon reservoirs situated near the beds laid down during the continental rift and Red Sea stages are not buried in the same places according to whether prospecting is undertaken on an American-type margin (Fig. 3.4) or on a European-type margin (Fig. 3.1). In the first case, productive drilling can be done on the present continental shelf. In the second case, drills have usually to be established at much greater depths (at the foot of the present continental slope). This greatly increases the technical difficulties.
3. Contourites are generally fine sediments (lutites), where each lamination is composed of well-calibrated sedimentary particles (the effect of sorting by the currents). The grain-size sorting is normal or reversed in different cases (showing the effects of reductions or increases in current velocities). The sedimentary material in the contourites is probably brought to the ocean-floor by turbidity currents. But this material is immediately taken up again by the deep currents. They spread it about in directions which are often very different from those taken by the turbidity currents. The continental rise is the area where contourite deposition most often occurs.
4. The evaporites at the base of the sedimentary sequences on the stable margins, like any other plastic layer intercalated in the sequence, can also help the cover become detached, slide and be deformed under the influence of gravity. 'Upslope', the extensional structures involved have been described in some cases (half-grabens or grabens filled with sediments as they formed). In contrast, the structures caused by the compression that should occur 'downslope' (i.e. at the foot of the slope on the continental rise) have not been clearly observed.
5. Transform directions are not necessarily the strict continuations of transform faults. The rotation pole of one plate in relation to another can be displaced as the opening occurs, which, of course, changes the direction of the active transform faults.
6. These slopes have been measured between -180 m and $-1\ 800$ m deep.
7. It will be recalled that two kinds of sediments take part in building the continental rise: *turbidites*, transported from the continent down the submarine canyons; and *contourites* resulting from the transport and deposition of sedimentary particles by deep currents.

Further reading

Burke C. A. and **Drake C. L.** (eds), 1974. *The Geology of Continental Margins*. Springer Verlag, New York, 1 009 pp.
Mascle J., 1976. 'Le Golfe de Guinée (Atlantique Sud): un exemple d'évolution de marges atlantiques en cisaillement', *Mém. Soc. Géol. Fr.* (nouvelle série), **55** no. 128, pp. 1–104.
Talwani M., Hay W. and **Ryan W. B. F.** (eds), 1979. 'Deep drilling results in the Atlantic Ocean: continental margins and paleoenvironment', *Maurice Ewing Series 3*. Amer. Geophys. Union, Washington DC, 437 pp.
Vanney J. R. 1977. *Géomorphologie des plates-formes continentales*. Doin, Paris, 300 pp.
Watkins J. S., Montadert L. and **Wood-Dickerson P.** (eds), 1979. 'Geological and geophysical investigations of continental margins', *Amer. Ass. of Petr. Geol., Mem. 29*. Tulsa, USA, 479 pp.

The structural and sedimentary evolution of stable margins

'Continental Margins of Atlantic type', *Proc. of an International Symposium,* São Paulo, 1976. *Anals da Academia brasilera de Ciëcias*, vol. 48, suppl.
'Histoire structurale du golfe de Gascogne', Symposium organisé par l'I.F.P. et le C.N.E.X.O., Rueil-Malmaison 14–16 décembre, 1970, *Coll. Colloques et Séminaires*, no. 22, 1971, 2 vol. Technip, Paris.

Chapter 4
The morphological and structural effects of subduction

The beginning of this chapter (§1) describes the main morphological and structural features developing on the boundary between two convergent plates: the outer swell on the descending plate, the ocean trench, the tectonic accretionary prism and the fore-arc basin, the volcanic arc, the marginal basin and the back-arc basin. A brief study of the seismic and gravimetric effects of subduction is followed by the description of the deep structure of the active island arcs and margins (§2). Then the structural features in the tectonic accretionary prism (§3) and in the volcanic arc and back-arc basins (§4) are analysed. Finally sedimentation in cordillera-type margins is explained in terms of the special geometry of the descending plate (§5).

Lithospheric convergence brings about the superposition of two plates: one, the over-riding plate, remains on the earth's surface, whilst the other, the descending plate, plunges into the asthenosphere. The boundary between these plates coincides with *active continental margins* ('Cordillera-type' margins) when the over-riding lithosphere carries a continent with it (e.g. South America east of the Pacific). The boundary coincides with *active island arcs* when it is bordered by a festoon of islands and a 'marginal basin' with an oceanic substratum (many examples in the north and west of the Pacific Ocean, Fig. 4.1). In both cases, the 'under-thrusting' of the descending plate beneath the over-riding plate is called *subduction*.

   This simple model has been envisaged by geologists and seismologists who have analysed the travel times and frequencies of seismic waves travelling beneath active island arcs and margins. These two parameters provide valuable information about the physical state of the milieu crossed by seismic shocks. A rigid milieu (the lithosphere) transmits S-waves with relatively high frequencies with a greater velocity than a viscous milieu (the asthenosphere) which also 'attenuates' the shocks and only transmits the lower frequencies. With a world-wide network of observatories at their disposal, seismologists have thus been able to piece together the geometry of the lithospheric plates beneath the active island arcs and margins. They have also been able to show that the lithosphere here can plunge into the asthenosphere to varying depths (Fig. 4.2). Without any question this discovery enabled the plate theory to make great strides forward because it was then understood how and where the lithosphere was 'consumed' that had been 'created' on the axes of the active mid-ocean ridges. It therefore became possible to study the relative motion of the plates and to interpret the geodynamic features occurring on their boundaries.

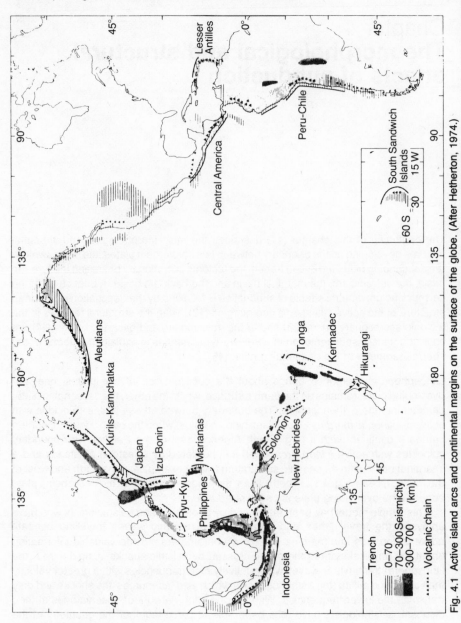

Fig. 4.1 Active island arcs and continental margins on the surface of the globe. (After Hetherton, 1974.)

1 The major morphostructural units

The major morphostructural units in an active island arc (and in an active cordillera-type margin without a marginal basin) are presented diagrammatically in Figure 4.3.

(*a*) The '*outer-swell*' probably results from a low widespread arching of the descending oceanic plate. It forms a wide arcuate swell that is about 200 km wide and rises some 200–400 m above the adjacent ocean floors. Its general form is probably

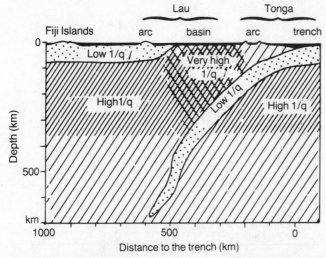

Fig. 4.2 Schematic cross-section of the lithosphere (dots) and the asthenosphere (diagonals) in the Tonga Islands area. $1/q$: attenuation factor (seismic waves are more attenuated as they are transmitted at lower frequencies). (After Barazangi and Isacks, 1971.)

caused by the bending of the descending plate and the weight of the over-riding plate (Fig. 4.4). The lithospheric zone on the convex side of the descending plate thus undergoes extension. The effects of this extension are seen on the surface as normal faults and fault-troughs separated by horsts. This deformation also causes shallow seismic activity, and some volcanic activity may occur locally.

(*b*) *The outer ocean trench slope* is relatively gentle (2% – 5% on average) and has the same structures *caused by extension* (and the same shallow seismic activity) as the outer swell. It, too, has horsts forming 'banks' and 'channels' along normal faults or grabens that run parallel or oblique to the trough axes.

(*c*) *The inner trench slope* is generally steeper than the outer slope (10% – 20% on average) and rises much higher and sometimes even reaches the surface. This gives the whole group an asymmetrical form (Fig. 4.5).

(*d*) *The ocean trench*. The morphological boundary line between two convergent plates is marked by the axis of the *ocean trench*. This is where the deepest soundings have been made (about 11 km in the Mariana trench). The trenches can, in some cases, have varying depths of infilling and they then have a flat floor caused by recent deposition.

The ocean trenches are about 100 km wide and about 1 000 km long : the Tonga–Kermadec trench extends 700 km whilst the Peru–Chile trench extends for 4 500 km. In active margins the trench axis roughly follows the fairly variable edge of the shelf (for example, in the Peru–Chile trench). This is a legacy of the past. On the other hand, the geometry of trenches associated with island arcs is generally determined by present or recent lithospheric movements. The over-riding plate is always on the concave side of the arc formed by the islands and the trench. The descending plate is always on the convex side (Fig. 1.14). As a result the boundary between two convergent plates (the curved axis of the trench) is rarely perpendicular to the direction of their relative motion. At the ends of the arc, the angle formed by this direction and the trench axis can even close completely. The boundary between the two plates then becomes a transform fault (Ch. 1).

The inner trench slope corresponds to one of the flanks of the tectonic *accretionary*

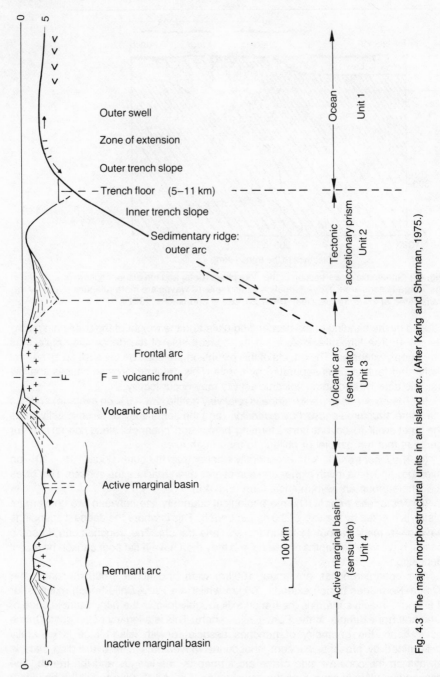

Fig. 4.3 The major morphostructural units in an island arc. (After Karig and Sharman, 1975.)

prism which will be studied in more detail later in this Chapter (*cf*. §3, p. 70). Within this zone, known as the 'subduction zone', the ocean sediments brought by the descending plate and the turbidites on the trench floor are deformed and partly incorporated into the over-riding plate. The growth of the prism causes the appearance of an '*outer*

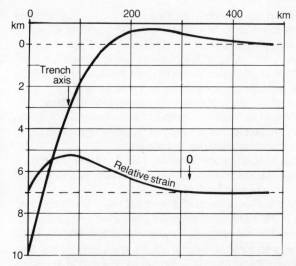

Fig. 4.4 Topographic profile nearing an ocean trench, calculated for a thin lithospheric plate. Forces causing extension appear in the upper part of the plate on the outer trench slope and the outer swell. (After Le Pichon *et al.*, 1973.)

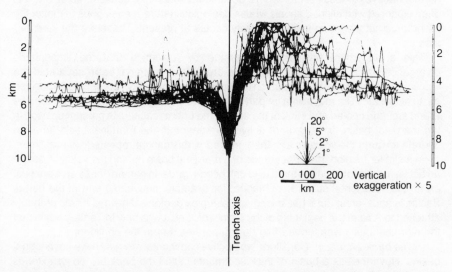

Fig. 4.5 Thirty projected bathymetric profiles of the Pacific island arcs aligned with respect to the trench axes. (After Hayes and Ewing, 1970.)

sedimentary ridge'. This can locally rise above sea-level (Barbados in the Lesser Antilles arc, Timor and Ceram in Indonesia, etc.) or, on the other hand, may merely be marked by a *trench slope break* high on the inner trench slope.

(*e*) *The fore-arc basin.* The sedimentary ridge can thus form a kind of dam behind which a *fore-arc basin* develops. There sediments accumulate consisting mainly of turbidites emanating generally from the volcanic arc. They are not deformed and not metamorphosed and are thus distinctly different from the dislocated formations in the accretionary prisms on which they lie unconformably.

(*f*) *The volcanic arc* (inner island arc) contains a *frontal arc* without present or recent vulcanicity, and a *volcanic chain* where, in contrast, varying numbers of active volcanoes occur in different cases. (In some arcs, as in the New Hebrides archipelago and the Tonga Islands a new sedimentary basin is inserted between these two areas.) Lastly the edge of the zone of active volcanoes facing towards the trench is called the '*volcanic front*'. (The older writers called this edge the 'andesite line'.)

Andesite vulcanicity is one of the most spectacular expressions of lithospheric convergence and the Pacific 'ring of fire' provides its most famous example. It will be described in Chapter 5 along with the plutonic activity that 'supplies' the island arc at depth and thus gives rise to continental crust out of the mantle.

(*g*) *The active marginal basin* is situated behind the island arc. Most often it is a zone of oceanic accretion where lithosphere is created by mechanisms comparable to those acting on the axes of active mid-ocean ridges (Ch. 1).

The initial cause of the formation and growth of a marginal basin is still not clear (possibly convection currents in the asthenosphere?). It is, however, certain that the process is related to subduction. The convergence between the descending plate and the island arc is thus associated with some 'drifting' of the island arc towards the open ocean. A plate limit ought logically therefore to be drawn along the axis of the active marginal basin where the oceanic lithosphere is created.

(*h*) *The marginal basin* sensu lato often comprises two sectors: (i) the active sector which has just been briefly described and is generally found near the volcanic arc, and (ii) the inactive, or fossil sector where oceanic accretion has ceased. Moreover, it is often cluttered with aligned shoals whose geological nature is analogous to that of the island arcs but where no volcanic activity occurs at present. These are the '*remnant arcs*'.

This arrangement has been interpreted by assuming that the lithospheric divergence which took place in the active area of the marginal basin separated the volcanic arc into two parts. The part on the ocean side would have preserved its position as an active arc. The much smaller part would have been gradually separated from the island arc, and cooled and sank at the same time until it reached its present position in the marginal basin. If the axis of active divergence in the marginal basin tends to migrate towards the volcanic arc, then, in theory, this can be repeated several times.

In a similar fashion, the appearance of a marginal basin behind the volcanic arc on an active margin must cause that arc to drift oceanwards. In general these arcs are built on an old continental substratum. Japan, for example, now forms part of the belt of Pacific Islands' arcs. But it has a long and complex geological history. It was probably attached to Asia at the beginning of the Cainozoic and it was only in Tertiary times that the marginal basin was formed that now separates it from the continent.

(*i*) *The back-arc basin.* Cordillera-type active continental margins have no marginal basins. Nevertheless a basin of thick sediments called the *back-arc basin* extends behind the frontal arc and the rising volcanic chain. Here thick volcanic and volcano–sedimentary sequences accumulate, as, for example, in the Andean Basin in Peru and Chile.[1] The subsidence in this basin may be explained by *crustal extension* which could correspond to the initial stage in the opening of a marginal basin. However, compressional regional deformation occurs on the border between the subsiding zone and the volcanic arc. This compression sometimes takes the form of an actual thrusting of the volcanic chain towards the continent (*cf.* (*c*), p. 75).

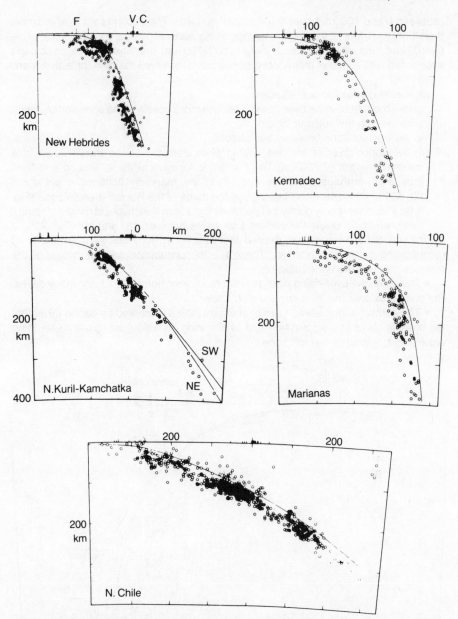

Fig. 4.6 Location of earthquake foci beneath active island arcs and margins. Most foci occur within a flattened mass called the Wadati–Benioff zone. The upper limit of this zone is indicated by the curved line on Fig. 4.7. (After Isacks and Barazangi, 1977.)

2 The deep structure

(a) The seismic effects of subduction

Earthquakes have a preferential distribution along two 'belts' (Fig. 1.10). The first belt coincides with divergent plate boundaries and only produces shallow earthquakes (between 0 and 100 km beneath the crustal surface). The second belt, in which more than three-quarters of the seismic energy of the earth is dissipated, includes *shallow* (0–100 km), *intermediate,* and *deep* (100–700 km) earthquakes. It is closely associated with zones of plate convergence (cordillera-type margins, or active island arcs).

Most earthquake foci are situated:

 (i) in the lithosphere in the over-riding plate and more exactly in the accretionary prism and beneath the volcanic arc;
 (ii) in the contact zone between two plates;
(iii) in the upper layer of the descending plate from the ocean trench down to a maximum depth of 700 km. The geometric location of these second and third groups of earthquake foci (ii) and (iii) is the markedly flattened mass about 20–30 km thick which has been given the name of the *Wadati–Benioff zone* (Fig. 4.6). This zone is only gently inclined near the ocean trench, but at greater depths it forms an angle below the surface that varies between 15° and 90° (Fig. 4.7).

The analysis of the signals registered after a tremor in many cases enables the 'focal mechanisms' to be reconstituted. These are the distribution and the nature of the stresses causing the earthquake:

 • The foci in the over-riding plate seem to result from normal faulting in many cases. But compressional mechanisms are also known.

 • The contact zone between two convergent plates is marked by earthquakes that are brought about by compression and by shearing (thrusting) along planes trending parallel to the Wadati–Benioff zone.

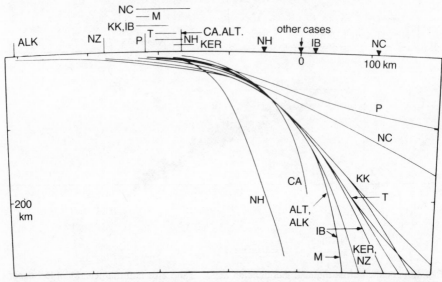

Fig. 4.7 The geometry of the Wadati–Benioff zone. The upper cover of this zone is represented by a different curved line for each example studied. NH: New Hebrides; CA: Central America; ALT: Aleutians; ALK: Alaska; M: Marianas; IB: Izu-Bonin; KER: Kermadec; NZ: New Zealand; T: Tonga; KK: Kuril-Kamchatka; NC: Northern Chile; P: Peru. (After Isacks and Barazangi, 1977.)

 • Most foci are situated in the upper layer of the descending plate (oceanic crust and mantle beneath the Moho). The focal mechanisms show that the forces causing the shocks are always aligned parallel to the direction of plate descent whatever the depth. But, in different cases, the earthquakes can result from either compression or extension. These features have given rise to many interpretations. Very recently, precise seismological studies have demonstrated that the Wadati–Benioff zone has really two parts. An initial plane, coinciding with the surface of the descending plate, is the location of the compressional earthquakes. A second plane is situated about 30 km beneath the first (and thus within the descending plate) and results from tensional stresses.

This unexpected distribution of earthquake foci and stresses has been interpreted in two ways:

(i) For some geophysicists, the upper plane corresponds to the friction surface where the descending plate rubs against the over-riding plate and against the asthenosphere immediately below (compressional forces). On the other hand, the earthquakes that are localised on the lower plane would be an effect of the gradual heating of the descending plate and of the stresses that such heating produces in an elastic milieu.

(ii) Other geophysicists explain the observed arrangement by evaluating the forces acting on the lithosphere when it first bends and then becomes a plane again at a depth of several tens of kilometres. Where the bend occurs tensional forces are created on the surface, whilst compressional forces develop lower down. But when the plate recovers its initial form, the places where the tensional forces have acted, in contrast now undergo the effects of compressional forces and vice versa, as if the lithospheric milieu acted in a plastic manner where it bent.

Three conclusions from these seismological studies should attract the geologist's attention:

(i) There is a zone near the contact surface between two plates that is affected by compression. (This will be considered again in Chapter 5 in connection with high pressure metamorphism).

(ii) The over-riding plate is deformed in the area between the volcanic arc and the ocean trench where either vertical or strike-slip movements occur.

(iii) The surface of the descending plate is subject to the structural effects of extension at the outer swell.

(b) The gravimetric effects of subduction and isostatic readjustments

The zones where plates converge always coincide with marked gravity anomalies which represent major zones of isostatic imbalance (Fig. 4.8).

• A slight *positive* anomaly (50 milligals) is associated with the outer swell and may be explained by the local rise in the sea-floor and the Moho through the bending of the lithosphere.

• A very marked *negative* anomaly may be observed on the ocean trench and the outer sedimentary arc (the accretionary prism). After all topographic corrections, this anomaly reaches or exceeds -200 milligals. It is clear that this important gravity deficit (which is, in fact, only 1/5000th of the total gravity) results from subduction. The shape adopted here by the descending plate, locally pushes the Moho and the mantle composed of dense rocks away from the surface. At the same time, on the other hand, the layer of water (in the trench) or sediment (in the accretionary prism), thickens considerably in these very same spots. These two features bring about a gravity deficit and in the end are the cause of the observed gravimetric anomaly.

As long as subduction lasts, the forces created by the plate convergence militate against isostatic readjustment. But if the subduction stops or slows down, then vertical movements (uplift) on a large scale can be expected to occur on the accretionary prism and the ocean trench. When lithospheric convergence ceases, therefore, a

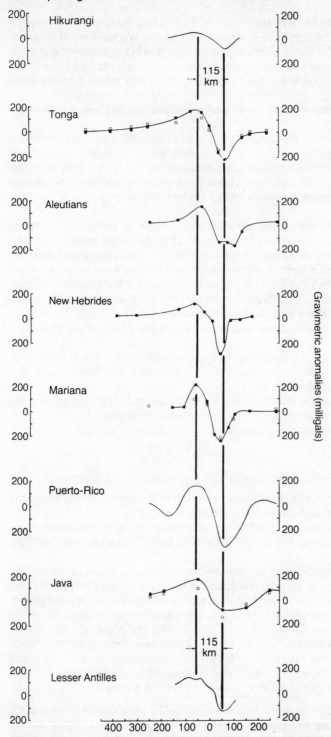

Fig. 4.8 Gravity anomalies on the borders of active island arcs and continental margins. The black dots represent free air anomalies, the white circles represent isostatic anomalies. Observations along profiles 1, 6 and 8 are dense enough not to be marked point by point (After Hatherton, 1974.)

sedimentary ridge is destined in principle to emerge and be eroded and then finally to reveal exposures of sediments that once were deeply buried. These processes imply the separation of the ridge and the volcanic arc, which usually occurs along major tectonic dislocations.

• On the other hand, the volcanic arc is characterised by a *positive* gravimetric anomaly. It is about 100–200 milligals so that the contrast between the gravity measured in the trench and that on the volcanic chain may exceed 400 milligals. The positive anomaly in the volcanic arc may be explained by the emplacement of plutons and surface volcanoes which are composed of materials emanating from the mantle (Ch. 5). Here, once again, is an effect of subduction and it should be diminished or subdued when the lithospheric convergence slows down or ceases entirely. For this reason, the volcanic arc is probably an area that is bound to subside if the margin or island arc to which it belongs stops being active.

In summary, the gravimetric data show that the subduction zones are in great *isostatic imbalance*. Under these conditions, when subduction slows down or stops, the return to equilibrium should be brought about by major vertical movements. Changes in subduction rates can, in fact, in many cases, be correlated with world-wide readjustments in plate movements. This synchronisation of tectonic phases (of extension or compression), whose effects can often be observed simultaneously in widely separated areas, can thus be explained.

(c) The nature and structure of the crust

The crust of the active island arcs and margins is very different in different zones:

1. The descending plate carries a typical oceanic crust

Subduction is theoretically possible only in this case. The relatively low density of a plate carrying continental crust, in principle, prevents it from sinking into the asthenosphere. In contrast, a plate covered with oceanic crust is denser than the asthenosphere because of its temperature. It can thus 'sink' into the asthenosphere under its own weight (Ch. 1).

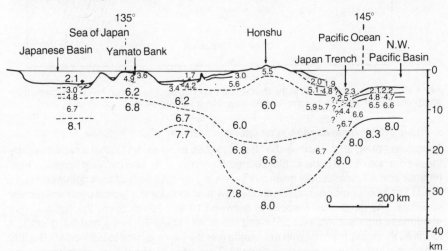

Fig. 4.9 Structure of the crust to the north-west of Japan. The numbers show P-wave velocities. (After Murauchi and Hasui, 1968, quoted by Sugimura and Uyeda, 1973.)

2. Some volcanic arcs are composed of typical continental crust

This is notably the case of all those arcs built up alongside a continent (cordillera-type active margins), or on a piece of continent that has been separated from its 'homeland' when a marginal basin opened (as in Japan for example) (Fig. 4.9). The continental structure is basically preserved in these conditions. The crust, however, is often enriched and *thickened* by plutonic and volcanic activity related to subduction (Ch. 5). Thus the 'roots' of the Andean Cordillera, and, as a result, its highest peaks, are situated more or less directly above the Neogene and recent volcanic arc (Fig. 4.10).

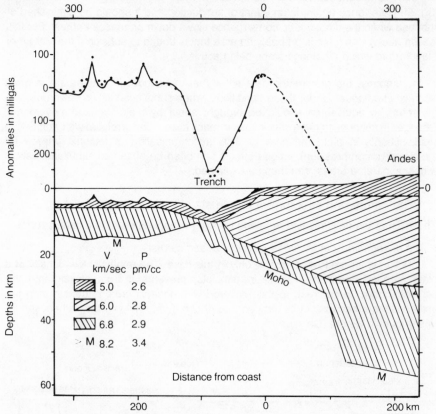

Fig. 4.10 Structure of the crust in the Peru-Chile trench area. The section has been drawn from interpretations of seismic and gravimetric data. (After Hayes, 1966.)

3. The crust of some volcanic arcs has 'intermediate' characteristics

These are young arcs, such as the New Hebrides arc (Fig. 4.11), that were formed by the subduction of one oceanic plate beneath another plate also carrying oceanic crust. Plutonic and volcanic activity gradually increased the thickness of the original crust and it thus lost its oceanic character without, however, acquiring a typical continental crust structure. The deep crust (especially between 10 and 30 km below the surface) has abnormally high seismic velocities and densities (about 7.5 km/s and 3.0 g/cm³). It forms such a small contrast with the mantle that the Moho is hard to distinguish. But this is probably only one stage in development. The zones of lithospheric convergence, as has been stated, are places where continental crust is made. The young arcs which are built on oceanic crust originally, are in the first stages of an evolution which, after a

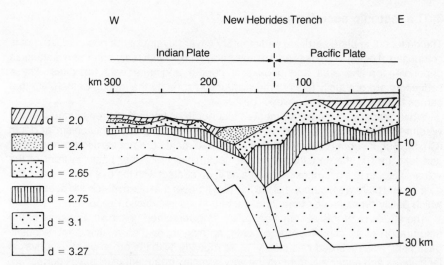

Fig. 4.11 Structure of the crust in the New Hebrides area. The section has been drawn from interpretations of gravimetric data. (After Collot and Misségué, 1977.)

prolonged period of subduction, should lead them to acquire all the characteristics of a continental crust.

4. The floors of large-scale marginal basins

The floors of large-scale marginal basins (a thousand kilometres or more across) are generally composed of typical ocean crust. The ophiolites in the fold mountains that most authorities associate with former oceanic crusts, could therefore come either from a large ocean or from a marginal basin. Their geotectonic and palaeogeographic significance would be quite different in the two cases.

The creation and evolution of a marginal basin or a large ocean probably obey similar laws. The 'thinned continental crust' seen at the foot of stable margins (Ch. 2) should therefore also exist in the marginal basins, especially at their inner edge (the side opposite the volcanic chain). It will be recalled, on the other hand, that some marginal basins have scattering of 'remnant arcs' with a continental substratum. This marks an important difference from the large oceans. Either because of their smaller size, or because of the presence of these remnant arcs, these marginal basins often display remnants of continental crust 'lost' in oceanic crust. This is not seen in 'real' oceans. The origin of this contrast is to be sought more in the age of the zone of oceanic accretion (the marginal basins are young oceans) than in the geodynamic processes causing this accretion. These processes are probably identical in both cases.

In summary, the trench–arc–marginal basin system is composed of zones that are differentiated fundamentally by the nature of their crust:

• In general the descending plate carries a typical *oceanic crust* on its surface.

• The volcanic arc can be composed either of *old continental crust* that has been enriched by contemporary plutonic and volcanic activity, or of an *oceanic substratum* that thickens gradually as the plutons and volcanoes form, and thus takes on *intermediate characteristics* between the continental and oceanic crusts.

• The floors of the marginal basins are often composed of *oceanic crust*, but they also reveal *relics of continental or intermediate crust*.

3 The tectonic accretionary prism

The descending plate carries a sedimentary cover (layer 1 of the ocean crust) that is composed of pelagic deposits (often brown clays mixed with radiolarites and diatomites depending on the area and geological period) and infrequent turbidites. These sediments are gradually led into the subduction zone by the motion of the plate that carries them along like a 'conveyor belt'.[2]

What happens to these sediments? It is not impossible that they plunge into the Wadati–Benioff zone and feed the vulcanicity and calc-alkaline plutonism at depth (Ch. 5). But, it seems likely that most of the sediments are not carried down deeply[3] and take part in the construction of the tectonic accretionary prism in front of the volcanic arc. In this hypothesis it seems as if, on contact with the over-riding plate, the descending plate were stripped of the thin plastic layer that blankets it in the oceans, but which easily breaks away from its basement in the subduction zone.

The structure of this accretionary prism long remained unknown. Because of the deformation of its constituent sediments, nothing more than a diffractant 'acoustic basement' was recorded seismically. More recently, with the progress of geophysical techniques and especially through the very complex treatment of seismic information, the structure of the prism has now appeared in the records. The picture obtained is one of a pile of flat wedges separated by abnormal contacts. These approach the horizontal near the base of the structure (i.e. at the foot of the inner trench wall), but have a greater dip towards the top of the structure.

(a) The 'model' of tectonic accretion (Fig. 4.12)

On seismic registers, the roof of the ocean basement (layer 2) forms a clear reflecting surface which can sometimes be traced for great distances. This is where the true boundary between two convergent plates is to be found and, happily, the direct observations of its dip verifies the existence of the phenomenon of subduction. However, some geological structures that are thought to be relict accretionary prisms contain ophiolite fragments and wedges. These suggest that, like the sediments, the oceanic crust can help feed the accretionary prism by scraping 'slivers' from the descending plate.

● According to the model in Fig. 4.12(a), the bulk of the sediments in the accretionary prism are of two types. Some sediments come from the cover on the ocean basement situated beyond the trench (deep pelagic deposits and infrequent turbidites). Other sediments are the turbidites accumulated in the ocean trench most often as a result of slumping on the steep inner slope (Fig. 4.13). In this case the sediments are produced either by reworking in the accretionary prism itself (which thus undergoes a kind of submarine erosion as it is being formed), or by sediments that are first deposited unconformably on the accretionary prism and then slump downslope and accumulate in the bottom of the trench.

● The youngest thrust planes are also nearest the horizontal and intersect the roof of the oceanic basement at a very acute angle. Everything suggests that the convergence of the plates forms a narrow sedimentary wedge, then a second and then a third, etc. at the base of the inner trench slope. Eventually a kind of fan is built up as each wedge is raised and tilted up as new tectonic units are added. Thus the thrust planes increase in age and dip from the base of the inner trench slope upwards.

● In the same fashion, the deformed sediments decrease in age towards the base of the inner trench slope, i.e. where the new tectonic wedges are emplaced. (In Figure 4.12(b), the conventional isochron lines are represented by the numbers.)

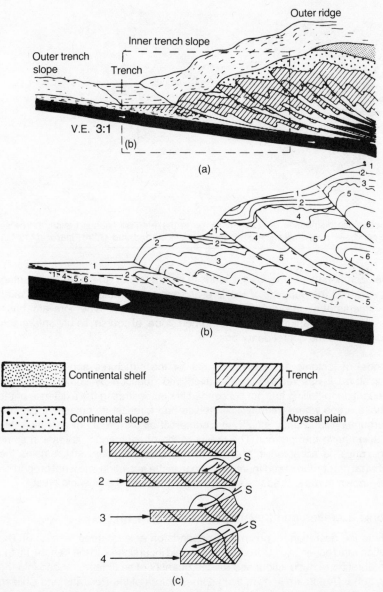

Fig. 4.12 Model of tectonic accretion. (a) distribution of facies; (b) distribution of isochrons (stratigraphy). The numbered sequences go from the youngest to the oldest beds. The dotted line represents the boundary between the turbidites in the trench and the sediments in the abyssal plain. (c) interpretative diagram; S: slip-surface for the sediments slumping down the inner trench slope and thus feeding the turbidites in the trench; 1 to 4: successive stages of tectogenesis determined by plate convergence. (After Seely et al., 1974.)

• As the tectonic activity continues, unconformable sediments are deposited on the inner trench slope. These can be locally deformed when movement is renewed on the wedges in the accretionary prism itself. (In Figure 4.12(b) the unconformity is represented by a wavy line.)

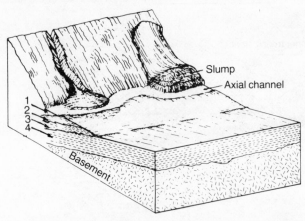

Fig. 4.13 Schematic distribution of sediment facies in the Aleutian Islands trench. 1: coarse turbidites; 2: sand, silt, mud; 3: silt and mud; 4: abyssal plain: silt and mud. (After Piper *et al*., 1973.)

● As the accretionary prism grows, it tends to form a barrier (the outer sedimentary ridge) above which the deposits of the fore-arc basin accumulate. These are often detrital sediments that are rich in volcanoclastic products. The fore-arc basin sediments, like those 'blanketing' the inner trench slope, of course, lie unconformably on the sediments in the accretionary prism.

This model of tectonic accretion accounts for the structures appearing on the seismic registers. It has recently been called into question by some authorities, because recent deep drilling has not succeeded in demonstrating the existence of the prism in those areas where geophysical studies had forecast its presence. But this negative argument can hardly effectively counteract all the proofs of accretion that seismic studies have brought out. The reality of the phenomenon, at least in some subduction zones, is accepted in the rest of this chapter. There still remains the problem of what proportion of sediments takes part in the accretion and what proportion are dragged down into the Wadati–Benioff zone along with the oceanic crust.

(*b*) Tectonic accretion and trench migration (Fig. 4.14)

The supply to the accretionary prism thus depends on at least three factors: (i) The sedimentation rate (especially the frequency of turbid incursions into the trench); (ii) the speed and duration of subduction; and (iii) the quantity of sediment dragged into the subduction zone. (Areas are known, as in some sectors of the Peruvian and Chilean Andes, where prolonged convergence has given rise only to an insignificant prism or to no accretionary prism at all.)

Figure 4.14 illustrates the hypothesis whereby continued tectonic accretion can cause the trench to migrate towards the ocean. This therefore increases the space between the trench and the volcanic arc and also widens the fore-arc basin. The weight of the accretionary prism brings about a local depression of the lithosphere just like any other comparable load. (Chapter 1, Fig. 1.7). But, in this case the trench only represents the front of the accretionary prism as it advances over the descending plate. It no longer represents the surface expression of the zone where the descending plate bends and penetrates into the asthenosphere. This zone is henceforth buried beneath the accretionary prism itself.

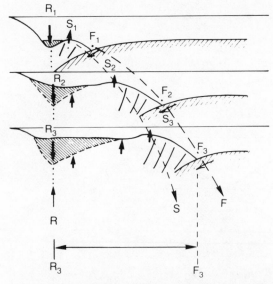

Fig. 4.14 The migration of the ocean trench through tectonic accretion. $F_{1, 2, 3}$: successive trench positions; $R_{1, 2, 3}$: fixed marker representing the trench floor when subduction starts (for example, the foot of the continental slope on a stable continental margin just before it becomes active); $S_{1, 2, 3}$: successive positions of outer trench slope break. The unconformable sediments in the fore-arc basin are represented by close diagonals, and the oceanic crust of the descending plate by wide diagonals. The extent of oceanward trench migration is measured by the distance $R_3 - F_3$. (Based on Karig and Sharman, 1975.)

4 The deformation of the over-riding plate

It will be recalled (*cf.* (*b*) p. 65) that active island arcs and margins are in isostatic imbalance and that changes in subduction rates (the relative speeds of convergent plates) probably give rise to considerable vertical movements within the over-riding plate. On the other hand, the ageing of the newly-formed lithosphere in the marginal basin causes it to subside like other oceanic areas (Ch. 2), and remnant arcs themselves sink as they cool (*cf.* (*h*), p. 62). But the over-riding plate also undergoes deformation generated by tangential forces.

(*a*) Synthetic and antithetic thrusts (Fig. 4.15)

A theoretical model of the deformation of the edge of the over-riding plate is suggested by Figure 4.15. It is based on the principle that a given shear plane, C_1, can be linked with another conjugate shear plane C_2. The compressional force F is then orientated along the bisector of the small angle formed by the intersection of C_1 and C_2. In the case considered (subduction), the plane C_1 represents the upper Wadati–Benioff zone (up to 100 km deep) and the corresponding thrusts are called *synthetic*, whilst the conjugate thrusts C_2 are called *antithetic* in relation to the subduction plane.

  • If the Wadati–Benioff zone is very gently inclined (which is most often the case, see Fig. 4.7), then the C_1 *synthetic* thrusts are facilitated (Fig. 4.15(a)) whilst the C_2 shears are of limited size and only give rise to regional uplift. The best example of synthetic thrusts is provided by the tectonic accretionary prism. The pitching occurs on the trench side (*pointing towards the ocean*) and the descending lithosphere can theoretically be

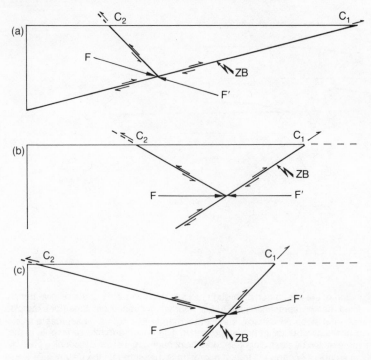

Fig. 4.15 Diagram of shearing in the over-riding plate in relation to the dip of the Wadati–Benioff zone Z.B. C_1: synthetic thrusts; C_2: antithetic thrusts.

'scraped', and thus take part in the formation of tectonic elements which then contain ophiolite.

● In contrast, when the Wadati–Benioff zone is steeply inclined (Fig. 4.15(c)) the synthetic shear plane steepens and the antithetic thrusts C_2 increase in importance. In principle, the wedging involves a vast area of the over-riding plate (including all the volcanic arc) where the crust is generally continental or intermediate in nature, but not, in any case, oceanic. Here the tectonic elements that are pitched towards the foreland or the marginal basin (*pointing towards the continent*) contain *no ophiolites.* It seems that this occurs less frequently than the preceding case.

● Finally, if the forces F which are responsible for the shears C_1 and C_2 are orientated parallel to the crustal surface (Fig. 4.15(b)), both the antithetic and the synthetic thrusts are of similar size and double pitching structures may result.

It will be seen that, according to this model, the tectonic style of lithospheric convergence zones is partly determined by variations in the inclination of the Wadati–Benioff zone.

(*b*) Transcurrent faults parallel to plate boundaries (Fig. 4.16).

The convergent motion between two plates is rarely perpendicular to their boundaries. When the relative displacement is markedly oblique to the trench, it may be anticipated that tangential forces operating on the over-riding plate will there give rise to transcurrent faults trending parallel to the trench axis. This transcurrent movement is either dextral or sinistral depending on the angle between the direction of relative motion and the plate boundary. In Fig. 4.16(a) and (b), situation (a) describes Japan, which is crossed by spectacular dextral transcurrent faults from one end to the other.

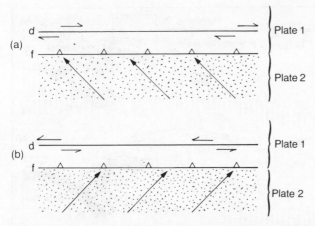

Fig. 4.16 Diagram of the transcurrent faults d induced in the over-riding plate by a lithospheric convergence trending obliquely to the boundary f between plates 1 and 2.

(c) The movements of the volcanic arc in relation to the back-arc basin (active continental margins of Cordillera-type)

The lithosphere is modified behind the subduction zone beneath the volcanic chain. The zone of plutons and volcanoes forms a sort of curtain where the rise of magmas weakens the plate. It is as if the volcanic arc separated two 'sub-plates' L_1 and L_2 (Fig. 4.17(a) and (b)) which can tend either to converge or to diverge.

• If extension predominates, i.e. if the volcanic arc tends to move away from the stable continent, then the volcanic chain is separated from the back-arc basin by normal faults (Fig. 4.17(a)). The back-arc basin can be thought of as the 'precursor' of a marginal basin, exactly as a continental rift precedes the opening of an ocean basin.

• If, in contrast, convergence occurs (Fig. 4.17(b)), then the back-arc basin is over-ridden by the volcanic arc. Some geophysicists have suggested that this movement could be interpreted as the beginning of the subduction of basin L_1 beneath arc L_2.

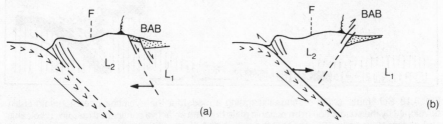

Fig. 4.17 (a) Extension between the volcanic arc and the back-arc basin BAB. The 'under-plates' L_1 and L_2 are diverging. F: volcanic front. (b) Back arc basin over-ridden by the volcanic arc. The 'under-plates' L_1 and L_2 are converging.

The size of these 'convergent' and 'divergent' movements is not, however, on the same scale as the subduction in front of the volcanic arc and in the accretionary prism. In particular, little difference is seen between the 'antithetic' thrusts studied by geologists on the boundary between the volcanic chain and the back-arc, and the 'subduction' suggested by geophysicists beneath the volcanic arc in this same basin. It is, in fact, the same feature which is described with a different vocabulary.

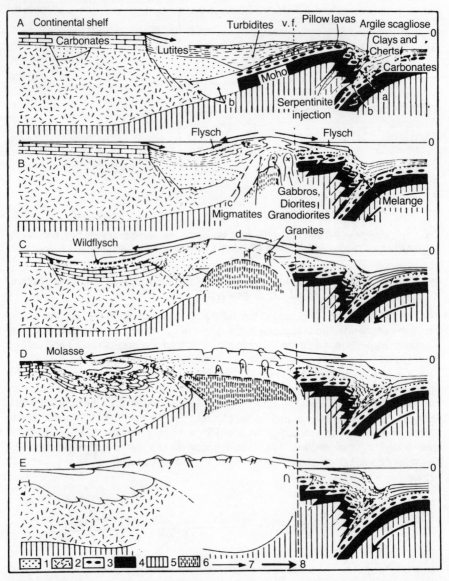

Fig. 4.18 Schematic cross-sections illustrating a model for the evolution of a mountain chain developed by the subduction of an oceanic plate beneath an active continental margin. 1: volcanic and sedimentary rocks; 2: continental crust; 3 and 4: oceanic crust; 5: mantle; 6: magmatic intrusion; 7: sediment transport; 8: plate movement; a: blueschist metamorphism; b: intermediate crust; c: metamorphic front; d: thermal doming. (After Dewey and Bird, 1970.)

● Figure 4.18 shows a model of the evolutionary sequence of a cordillera-type active margin. In this model the *antithetic thrusts* (the volcanic arc thrust over the 'back-arc basin') occur *later than the extension behind the arc* (infilling of back-arc basin by thick volcanic and sedimentary sequences). This is what geologists have observed in the Andean Cordillera where the phases of squeezing, when they occur, always follow long periods of subsidence.

(*d*) The extension of island arcs (Fig. 4.19).

The migration of the island arc during the growth of the marginal basin (*cf*. (*g*), p. 62) determines the development of two groups of tectonic troughs.

• *Radial extension* gives rise to grabens trending parallel to the volcanic chain.

• The curvature of the boundary between the two convergent plates and the oceanward bulge of the arc imply that the leading edge of the over-riding plate also undergoes *longitudinal stretching* which determines the development of tectonic troughs at right angles to those already described.

Whatever the orientation of the troughs may be, their sedimentary infilling (by subsidence) demonstrates successive stages of extension and thus also shows the gradual widening of the marginal basin behind the island arc.

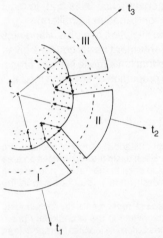

Fig. 4.19 Theoretical diagram of the fracture-pattern produced by radial and longitudinal extension in an arc. Dots: zones of extension; arrows show the size of the separating movements t_1, t_2, t_3 between the separated blocks I, II, III. This theoretical model was developed from the study of the Tyrrhenian arc in southern Italy. (After Dubois, 1976.)

5 Transverse structures on the active margins

The active margins have hitherto been studied along profiles trending at right angles to the axes of the trenches, as if the margins remained identical in whatever segment is considered. In fact *transverse structures* may be observed on the over-riding plate (often in bundles of faults), which divide the volcanic arc and the back-arc basin into *compartments*. The geological and geophysical effects of subduction are very different in different compartments (dip of the Wadati–Benioff zone, periodicity of earthquakes, intensity and chemistry of the vulcanicity, margin morphology, etc.).

These variations between compartments have a deep-seated origin. This is why the 'segmentation' of the active margins has been considered as an effect of the deformation of the descending plate. It is seen as being divided into a series of blocks like the variously depressed keys on a piano keyboard. Thus the contrasts that are seen on the over-riding plate in the intensity of subduction features may in some way be caused by sudden lateral variations in the dip of the Wadati–Benioff zone.

Whatever the reason, it is clear that the active continental margins (and the active island arcs as well) do not have constant characteristics throughout their lengths. The

Andean-type Cordilleras are no more 'cylindrical' structures than any other fold mountain chains.

In summary, this enumeration of the tectonic consequences of subduction brings out the fact that two sorts of structures arise when plates approach each other:

• *extension structures* orientated on axes trending parallel to the volcanic arc (back-arc basin, marginal basin) or at right angles to this arc (radial trenches);

• *compression structures* in the volcanic arc (transcurrent faults), in front of this arc (tectonic accretionary prism developing oceanward), or on its boundary with the back-arc basin, or the marginal basin (continentward overthrusts);

Both these types of structure may be explained by *horizontal* forces, but the subduction zones are also affected by great *vertical* movements. The volcanic arc, especially, is greatly uplifted as long as lithospheric convergence lasts and, in most cases, forms vigorous relief features undergoing erosion (Andean Cordillera).

It must however be acknowledged that the structural evolution of active island arcs and margins is much less well known than that of the stable margins. Of course a given 'model' can put some sort of order into the observations that have made in these areas, but hypotheses and interpretations still take up a large proportion of any such model.

Notes

1. The Andean and Chilean back-arc basin is really discontinuous and a variety of sections are arranged 'en échelon' and trend obliquely to the ocean trench. Each basin thus comprises an area where sediments undergo strong oceanic influences on the side open to the Pacific, and an area (where the sediments have basically continental facies), which ends in a cul-de-sac in the continental interior.

2. This picture is not wholly correct in so far as it implies an 'absolute' movement of the descending plate towards the over-riding plate. Sediments may also accumulate in the accretionary prism from the gradual advance of the over-riding plate which pushes before it deposits 'scraped' from the ocean crust. What matters in the end therefore is the relative motion of the two plates.

3. The specialists are far from being in agreement on this point. At first this idea was rejected by geochemists who could not detect any evidence that these sediments took part in the creation of magma beneath the island arc. Some authorities now believe that layer 1 can plunge down with the descending plate and even that the over-riding lithosphere itself undergoes tectonic erosion in the friction zone between the plates. According to this view, the lithospheric matter lost in this way by the over-riding plate could contribute at depth towards the generation of the initial magmas erupted in the volcanic chain. In the present state of knowledge, however, this interpretation is still insecurely based. For this reason it has not been considered again in this chapter – nor in Chapter 5.

Further reading

Dickinson W. R. and **Seely D. R.**, 1979. 'Structure and stratigraphy of fore-arc regions', *Amer. Ass. of Petr. Geol. Bull.*, **63** (1) pp. 2–31.

Sigimura A. and **Uyeda S.** (eds), 1973. 'Island arcs: Japan and its environs', *Development in Geotectonics*, **3**. Elsevier Publishing Company, Amsterdam, 247 pp.

Sutton G. H., Manghani M. H., Moberly R. and **McAfee E. U.** (eds), 1976. *The Geophysics of the Pacific Ocean Basin and its Margin.* Amer. Geophys. Union, Washington, 1 vol., 480 pp.

Talwani M. and **Pitman W. C.** (eds), 1977. 'Island arcs, deep sea trenches and back-arc basins', *Maurice Ewing Series 1.* Amer. Geophys. Union, Washington DC, 470 pp.

Uyeda S. (ed.), 1979. 'Processes at subduction zones', *Tectonophysics*, **57** (1) pp. 1–94 (Special issue).

Watkins J. S., Montadert L. and **Wood-Dickerson P.** (eds), 1979. 'Geological and geophysical investigations of continental margins', *Amer. Ass. of Petr. Geol., Mem. 29*, Tulsa, USA, 479 pp.

'*Géodynamique du Sud-Ouest Pacifique*'. *International Symposium*, Nouméa, New Caledonia, 27 August–2 September. 1976. Technip, Paris, 413 pp.

Chapter 5
The magmatic and metamorphic effects of subduction

This chapter briefly describes the thermal and petrological features (§1) associated with subduction. The analysis of the lavas erupted by the volcanic chain enables their sources to be located in the mantle of the over-riding plate and in the oceanic crust of the descending plate, near the Wadati–Benioff zone, 80 km or more deep (§2). The distribution of pressures and temperatures in the lithosphere suggests that thermal metamorphism occurs directly beneath the volcanic arc, whilst high pressure metamorphism affects the ocean crust and sediments at depth in the zone of plate friction (§3).

1 Heat flow anomalies on active island arcs and margins

Heat flow generally remains remarkably constant on the surface of the globe (1.2 H.F.U.)[1] but registers great regional variations in the zones of lithospheric convergence. It drops markedly above the accretionary prism and the fore-arc basin (Fig. 5.1) to 0.5 to 0.8 H.F.U. In contrast, it reaches two or three times its normal value on the volcanic arc. Lastly, though it remains less than on the volcanic arc, it stays abnormally high in the active marginal basins.

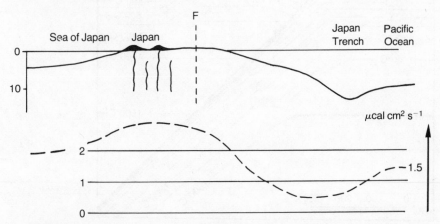

Fig. 5.1 Heat flow along a profile through Japan and the Japan trench. F: volcanic front. (After Dewey and Bird, 1970, simplified.)

• The positive thermal anomaly on the volcanic arcs is clearly related to their characteristic activity (magma and hydrothermal fluids are heat vectors).

• The negative anomaly associated with the subduction zone is interpreted as a result of the movement of the descending plate. Because of its thermal inertia, the lithosphere plunges down into the asthenosphere without immediately changing the geometry of the isotherms. Thus the geothermal gradient and the heat flow are considerably reduced at the tectonic accretionary prism and the fore-arc basin, because the surfaces of equal temperature descend with the plate carrying them (Fig. 5.2).

occurring at depth. Temperature, along with pressure, is one of the most important factors controlling petrogenesis. Because subduction has thermal consequences, its effects may be expected to be seen in rock transformation either in the solid state (metamorphism) or in the molten state (magmatism).

2 The magmatic effects of subduction

Active island arcs and margins are associated with intense volcanic activity which testifies to considerable magmatism. (The Pacific island arcs, for example, are emphasised by the 'ring of fire'). The volcanic arc is a zone which is being uplifted during subduction (Ch. 4) and erosion may reveal old plutons. These always appear closely related both geochemically and geometrically to volcanoes of the same age. Thus, what is described in this paragraph for vulcanicity is also valid for the phenomenon of magmatism in general. Thus the volcanic association basalt ($SiO_2 < 53\%$) – intermediate lavas – rhyolites ($SiO_2 > 68\%$) has also an equivalent plutonic association.

The volcanoes associated with present-day subduction zones generally emit lavas belonging to the *andesite* suite (basalts, greatly predominant andesites, dacites, and rhyolites). This type of vulcanicity is often explosive because the lavas are rich in gases

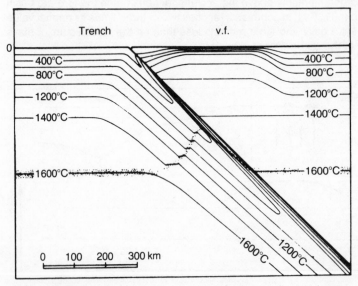

Fig. 5.2 Thermal structure of a lithospheric convergence zone calculated from surface heat flow values (no vertical exaggeration), v.f: volcanic front.

and especially in water-vapour. To this may be added the dangers of intense seismic activity and the resulting tidal waves. It will therefore be appreciated how inhospitable and prone to frequent natural catastrophes active island arcs and margins really are.

(a) Geochemical characteristics of volcanic rocks

It will be recalled that the geochemical characteristics of a volcanic rock depend on at least two factors:

• The composition of the initial magma ('the primary magma') which is determined (i) by the chemical and mineralogical composition of the *'parent rocks'* which give birth; (ii) by the temperature and pressure conditions that bring about their *partial melting*. The least refractory minerals are the first to contribute to the formation of the liquid phase. This causes a geochemical fractionation between the primary magma and the solid residue.

• The differentiation of this magma by *fractional crystallisation* of the minerals as the temperature and pressure decrease. The first minerals to form from the liquid are those with the highest melting point.

The geochemical classification of the volcanic rocks emitted on the active arcs and margins should therefore be 'genetic'. It should, in other words, separate lavas emanating from different primary magmas and 'parent rocks' or produced by different processes of differentiation. Three 'series' can be distinguished especially according to their potassium (K_2O) and iron oxide content.

1. The tholeiitic series

The tholeiitic series is characteristically poor in K_2O (less than 1%, see Fig. 5.3) and in TiO_2. It is also characterised by iron enrichment and increase in the FeO/MgO ratio during differentiation and by the consistently high value of this ratio (Fig. 5.4). The series consists especially of *basalts* (very similar to the rocks comprising the oceanic basement) and icelandites (andesite lavas poor in potash) and infrequent dacites.

• The iron enrichment in relation to magnesium is explained by the crystallisation of olivines and pyroxenes in which the ratio Fe/Mg is below that of the residual liquid.

• The low alkaline content ($K_2O + Na_2O$) is probably attributable to the composition of the primary magma, (which is itself poor in these elements) and to the low differentiation of this magma.

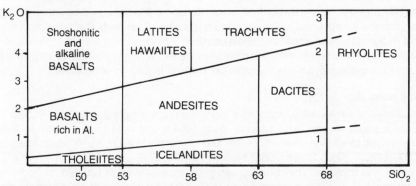

Fig. 5.3 Schematic chemical classification of the volcanic rocks on active island arcs and margins. 1: tholeiitic series; 2: calc-alkaline series; 3: potassium series. (After Barberi, 1974, simplified and modified.)

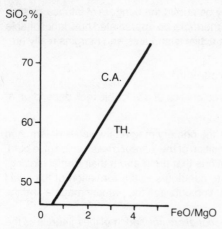

Fig. 5.4 Division of tholeiitic and calc-alkaline volcanic rocks according to their SiO_2 content and the FeO/MgO ratio. (After Miyashiro, 1974.)

The volcanoes erupting rocks in the tholeiitic series are closest to the ocean trench and generally mark the 'volcanic front' (Fig. 4.1 and Fig. 5.5). They are absent from old active arcs and margins (i.e. those that have functioned a long time), as in Peru. But in contrast, they are well represented in young arcs (at the start of their evolution).

2. The Calc-alkaline series

The Calc-alkaline series is most characteristic of the zones of lithospheric convergence. Compared with the tholeiitic series, this series is richer in K_2O (Fig. 5.3) and in TiO_2. (The andesites of the active island arcs are, however, poorer in potassium than those of the active margins). It has, moreover, a low iron content (Fig. 5.4) and contains, on the other hand, a high proportion of alumina. For example, the calc-alkaline andesites contain 16–18 per cent Al_2O_3 whilst the icelandites in the tholeiitic series only have 13–14 per cent. There are in the calc-alkaline series some basalts that are rich in alumina ($SiO_2 < 53\%$), *andesites*, which are by far the most frequent lavas ($53\% < SiO_2 < 63\%$), dacites ($63\% < SiO_2 < 68\%$) and rhyolites ($SiO_2 < 68\%$) which are sometimes associated with great ignimbrite masses. (The corresponding plutonic association is formed by the gabbro-diorite-granodiorite-granite 'series').

● The relatively low iron content which does not increase in proportion during differentiation is explained by the early crystallisation of minerals that are rich in this metal (magnetite, ferriferous amphibole, etc.).

● The slight enrichment in K_2O in relation to the tholeiites may be interpreted as the result of the greater pyroxene crystallisation (which is poor in potassium).

3. The potassium series

The potassium series lastly, contains shoshonite lavas, where the K_2O/Na_2O ratio is above or equal to one, and alkaline lavas where the ratio is less than 1. It is characterised by its high alkaline content and especially in K_2O (> 2–4% according to its SiO_2 content, Fig. 5.3). It includes basalts, hawaiites (alkaline), latites (shoshonitic), trachytes and rhyolites. The corresponding volcanoes are often situated far from the ocean trench – sometimes as much as 200–300 km away, in the areas where the Wadati–Benioff zone is very deep (Fig. 5.5). But alkaline rhyolites and trachytes can also be found in any part of the volcanic chain. As a result, some authorities believe that

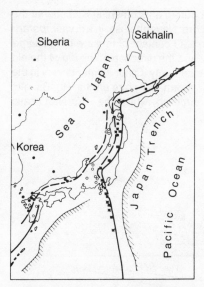

* Tholeiitic series   ○ Calc-alkaline series

● Alkaline series

Fig. 5.5 Distribution of the volcanic series in Japan. Continuous line: volcanic front; dashed lines: boundaries between the zones of tholeiitic volcanoes (stars), calc-alkaline volcanoes (open circles), and alkaline volcanoes (dots). (After Kuno, quoted by Lefevre *et al.*, 1974.)

the generation of this type of magma is not directly related to subduction, and that their deep origin should not be sought within the Wadati–Benioff zone which is where the tholeiitic and calc-alkaline series are created (*cf.* (*d*), p. 85). Whatever the reason, the lavas of the potassium series (shoshonites or alkaline) are only abundant on old active arcs and margins that have been evolving for a long time.

(*b*) The relationships between vulcanicity and the depth of the Wadati–Benioff zone.

The volcanic front is usually situated some 80 or 100 km above the Wadati–Benioff zone and the majority of volcanoes are between 100 and 150 km above it. But it has already been noted that emissions of alkaline rocks occasionally occur even further from the volcanic front.

The chemistry of the lavas (in the tholeiitic and calc-alkaline series) seems partly to depend on the distance from the trench, i.e. on the depth of the seismic zone beneath the volcano. The K_2O content, especially, increases regularly from the volcanic front away from the trench (Fig. 5.6). These, of course, are tendencies with local exceptions. But it is a general feature which suggests that the origin of the magmas should be sought somewhere near the seismic zone because the depth of this zone influences the composition of the lavas.

These geochemical 'gradients' are not only valuable in clarifying the deep geodynamic phenomenon which lies at the origin of magmatism (*cf.* (*e*), p. 87). They also enable the direction of 'Palaeo-subduction' to be re-established in remnant volcanic arcs when there is no longer any seismicity to provide any evidence about the geometry of the Wadati–Benioff zone. A precise quantitative relationship has even been suggested between the depth of this zone beneath the volcano and the potassium content. But the ageing of an arc often brings about an enrichment in K_2O so that the potassium content really depends on several factors. Nevertheless, in a *contemporary* calc-alkaline magmatic series, a variation in the K_2O content of rocks with the *same quantity of silica* enables the *trench-arc polarity* and the direction of the dip of the old Wadati–Benioff zone to be established with some certainty. Other elements vary in the same way as potassium and thus provide similar information: rare earths, Rb, Ba, etc. In the same way FeO/MgO and K/Rb ratios diminish, whilst the K_2O/Na_2O ratio increases with distance from the volcanic front.

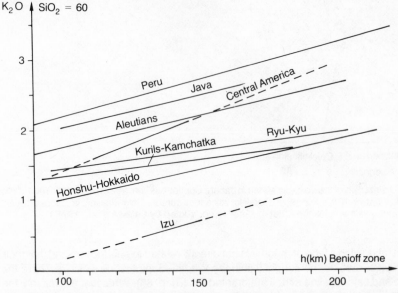

Fig. 5.6 K_2O content in andesites (for SiO_2: 60%) in relation to the depth of the Wadati–Benioff zone in different active island arcs and margins. (After Nielson and Stoiber, 1973.)

It must however be noted that, though the source area of the magmas is to be sought near the seismic zone, the two are not necessarily the same. It will be recalled (Ch. 4), that the most frequent foci are situated *within the descending plate*. Now, the rise in temperature and the liberation of water-vapour (*cf.* (*d*), p. 85) that are probably responsible for partial melting and the creation of primary magmas, ought to occur very close to the *surface of the descending plate*. Thus the magma source areas and the true Wadati–Benioff zone are probably separated by a certain distance.

(c) The possible migration of the volcanic front

The gap between the ocean trench and the volcanic front is markedly wider (200–300 km) on the older active island arcs and margins that have functioned for a long time, than on the zones of recent lithospheric convergence (100–200 km). More exactly, this gap (which includes the fore-arc basin and the tectonic accretionary prism), seems to increase in proportion to the length of time subduction has occurred (Fig. 5.7). This evolution probably springs from two distinct features:

• The 'supply' to the accretionary prism which causes the trench to migrate oceanwards (*cf.* (*b*), p. 72 and Fig. 4.14).

• The opposing ('continentward') movement of the volcanic front. This is most often explained by a progressive lowering of the isotherms under the over-riding plate after a long period of subduction. In this case, the zone of partial melting where the magmas are created also deepens and shifts towards the continent.

As a result of this migration, at least part of the frontal arc (which has no volcanic activity at present), can nevertheless contain old intrusions or extrusions that were emplaced during a previous stage of subduction. This may be observed especially in the Andean Cordillera (in the Peru and north Chile area). Here the gradual shift of the volcanic chain towards the continent during the Mesozoic and Cainozoic has also caused the back-arc to migrate in the same direction.

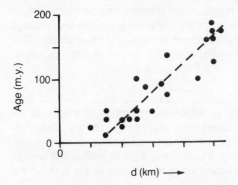

Fig. 5.7 The gap between the volcanic front and the axis of the ocean trench in relation to the age (in millions of years) of the oldest magmatic activity observed in about twenty active island arcs and margins. (After Dickinson, 1973, simplified.)

(d) The origin of magmas

The old hypothesis that the main source of magmas was within the continental crust of the volcanic arcs, can now be abandoned forever as a result of many petrological and geochemical arguments (notably from the study of the initial isotopic ratios of strontium). For the same reasons, most petrologists think it unlikely that the sediments that are dragged into the subduction zone make any effective contribution to magma formation (except, perhaps by bringing in water-vapour). Even if, in certain cases, the continental crust or these sediments cause any contamination, the main origin of plutons and volcanoes should still rather be sought either *in the oceanic crust in the descending plate, or in the mantle in the over-riding plate*.

Experimental petrological studies have shown that several kinds of rock can produce tholeiitic or calc-alkaline magmas by partial melting comparable to those observed in the volcanic arcs.

Ultramafic rocks of the upper mantle

Experimental study of the partial melting of lherzolites (rocks from the upper mantle) has shown that, in the presence of water-vapour, they give rise to liquids saturated in silica when subjected to the temperature and pressure conditions thought to exist at depths of 70–100 km (approximately under the volcanic front). When the total pressure and the partial water-vapour pressure are reduced, olivine (+ pyroxene) crystals are formed and the residual liquid gradually takes on an andesite composition. This differentiation would occur at low pressures, about thirty kilometres down and the evolution of the primary magma would then create rocks in the *tholeiitic* series.

Amphibolites

Experimentally, the partial melting of amphibolites begins, *under an important partial water-vapour pressure* (3–5 Kb), in the pressure and temperature conditions achieved at shallow depths (30–60 km). It will be recalled that the volcanic front is situated 80–100 km above the seismic zone near which the magmas are believed to be created. Amphibolites cannot therefore be considered as a possible source of the plutons and volcanoes in the volcanic arc.

Nevertheless, the oceanic crust of the descending plate, which has undergone initial thermal metamorphism, probably contains important quantities of amphibolites. As these make no direct contribution to magmatism in the subduction zones, it must be supposed that the water-vapour pressure is insufficient to cause their partial melting. In this case, amphibolites subject to increases of temperature and pressure undergo *metamorphism* in the solid state that transforms them into quartz-eclogites. This reaction liberates a certain amount of water-vapour (amphibole $\longrightarrow$ garnet + pyroxene + H_2O).

Quartz-eclogites

These rocks of basaltic composition begin to melt *in the presence of water-vapour* when subjected to the pressures and temperatures that occur naturally about 100 kilometres deep. The liquids so formed are poor in iron and rich in potassium (in relation to the initial rock). They first form dacites and rhyolites and then andesites and basalts as partial melting continues. The corresponding magmas can give rise to rocks in the calc-alkaline series by differentiation.

All the experimental petrological studies just briefly outlined have therefore shown that the presence of water-vapour is indispensable to the development of partial melting near the Wadati–Benioff zone either in the mantle in the over-riding plate (peridotites) or in the oceanic crust in the over-riding plate (quartz-eclogites). As the magmas are produced as long as the subduction lasts, this water must therefore be brought continuously to their source area. The water 'carriers' should be sought therefore in a lithospheric zone containing hydrated rocks or minerals that are continually renewed at depth. Only the oceanic crust in the descending plate fulfils these two requirements.

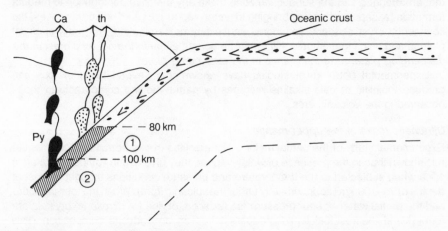

Fig. 5.8 Magma formation in subduction zones. 1 and 2: stages in formation described in the text; th: tholeiitic series; Ca: calc-alkaline series. (After Ringwood, 1974, simplified.)

• The hypothesis that water imbibed by sediments dragged into the subduction zone takes part in the process, is most unlikely at such a depth. On the other hand, the sediments do contain a small proportion of hydrated minerals (micas and amphiboles) which can liberate water-vapour in the subduction zone.

• The oceanic basement harbours considerable quantities of water fixed in its constituent rocks and minerals. It contains amphibolites and serpentinites especially, which are both rich in hydrated minerals. The amphibolites, as has been seen, are probably metamorphosed into quartz-eclogites by liberating water-vapour at a depth of about 100 kilometres. The serpentinites are altered at greater depths. Experimental studies have shown that they are dehydrated in several stages under high pressures, i.e. in subduction zones at levels between 100 km and 300 km below the surface.

Both these phenomena (amphibolite and then serpentinite dehydration), can thus create the conditions required (the presence of water-vapour) for the partial melting of quartz-eclogites and peridotites in the upper mantle.

(e) Petrographic synthesis

Petrologists have been able to make 'models' explaining the formation and evolution of the magmas emplaced in volcanic arcs, through experimental work and the formulation of hypotheses about the temperatures and pressures prevailing near the Wadati–Benioff zones. Figure 5.8 represents one such model.

Initial stage (young volcanic arcs, magmatism closest to volcanic front). At a depth of 80–100 km (temperatures about 650–700°C, pressure about 30–40 Kb), the amphibolites in the oceanic crust change in the solid state into quartz-eclogites + water-vapour.

The freed water-vapour brings about the partial melting of the upper mantle in the over-riding plate and the rise of magmas in diapirs. This relatively shallow magma differentiation, notably by the fractional crystallisation of olivine, gives rise to the *tholeiite series* observed in many arcs near their volcanic fronts.

Second stage (calc-alkaline magmatism). At greater depths (100–300 km) high water-vapour pressure is maintained on the surface of the descending plate because of the gradual dehydration of the serpentinites in the oceanic crust. The quartz-eclogites begin partial melting when the temperature rises above 750°, especially in the zone between 100 km and 150 km deep.

As the temperature and pressure decrease in shallower zones, these magmas become richer in potassium and poorer in iron through the crystallisation of pyroxene and garnet, and eventually give rise to rocks of the calc-alkaline series near the surface.

The problem of the potassium series

Alkaline rocks and shoshonites present a difficult petrological problem. They have, until recently, been considered as the last part of a continuous series that includes the tholeiite and calc-alkaline series. But many geochemical studies suggest that they probably do not come from the same source areas. However, no satisfactory 'model' has yet been proposed to explain their origin.

The causes of the spatial distribution of potassium

The increase in the K_2O content in the lavas as the volcanoes occur further from the volcanic front, has been explained by two distinct features which can, however, act at the same time:

(i) The gradual *partial melting* of the oceanic crust in the descending plate liberates potassium from potassium-rich minerals (phlogopite) as these are destroyed.

(ii) Magma differentiation under the volcanic arc by *fractional crystallisation* (e.g. Pyroxene creation can increase the K_2O content of the residual liquid).

(f) Conclusion: the formation of new continental crust

Even if some features have not yet been wholly clarified, the generation of magmas in zones of lithospheric convergence takes part in *the formation of continental crust out of the mantle*:

• either directly, by the partial melting of the peridotites in the over-riding plate,

• or indirectly, by the partial melting of the descending ocean crust, newly formed from the mantle on the axis of the mid-ocean ridges.

Conversely, beyond the depths where it gives rise to magmas, the descending plate carries refractory rocks (the residues of partial melting), which penetrate into the asthenosphere without being able to melt. The process therefore ends in an *irreversible differentiation*. Its most important result for the geologist is the thickening of the continental crust or even its renewal if the subduction occurs beneath an oceanic plate (Ch. 4). This conclusion is of considerable importance. Not only does plate tectonics explain the creation of oceanic crust in zones of lithospheric divergence, but it also explains *the formation of continental crust* in convergence zones. It must be said, however, that the result is less durable in the first case (the creation of new oceanic crust) than in the second case (the creation of new continental crust). This is because oceanic lithosphere can be consumed in subduction zones whilst continental lithosphere stays on the earth's surface as a rule.

3 The metamorphic effects of subduction

(a) Summary of the main types of regional metamorphism

Three major types of metamorphism, governed by temperature and pressure conditions when petrogenesis takes place, have been defined by comparing experimental petrological work and regional studies of metamorphic rocks:

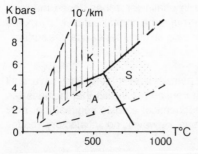

Fig. 5.9 Stability regions of the minerals characteristic of medium pressure metamorphism (K) and L.P.–H.T. metamorphism (A–S). K: Kyanite; S: sillimanite; A: andalusite. (After Miyashiro, 1972.)

Low pressure metamorphism

Low pressure metamorphism (L.P.) is characterised by andalusite which becomes an unstable mineral when the pressure exceeds 5Kb (Fig. 5.9). Under natural conditions, such a type of metamorphism can only develop in high temperatures. This implies a high regional geothermal gradient (more than 25° per km). In this case, it is called 'thermal metamorphism', or better, low pressure – high temperature metamorphism (L.P.–H.T.). The required heat can come simply from thermal conduction, but L.P.–H.T. metamorphism is most often related to the migration of magmas and fluids, which are important carriers of thermal energy. This type of metamorphic rock is thus often associated with granites or granodiorites.

High pressure metamorphism

High pressure metamorphism (H.P.) or blueschist facies metamorphism is characterised by glaucophane and jadeite, which are minerals that become unstable when the temperature exceeds 100°–300° according to the pressure (Fig. 5.10). High pressure metamorphism is therefore also a low temperature metamorphism (H.P.–L.T.). Under natural conditions it can only occur if the geothermal gradient is low (about 10° per km).

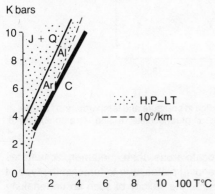

Fig. 5.10 Stability regions of minerals characteristic of H.P. – L.T. metamorphism. J: Jadeite; Q: Quartz; Al: Albite; Ar: Aragonite; C: Calcite. (After Miyashiro 1972.)

These two types of metamorphism occupy regions that are clearly separated and distant from each other in the field defined by temperatures and pressures. Between them, medium pressure metamorphism (or intermediate metamorphism) is characterised by kyanite and by the absence of any H.P.–L.T. minerals. The regional geothermal gradient for this type of metamorphism is about 25° per km.

(b) The distribution of temperatures and pressures in lithospheric convergence zones and the metamorphic consequences

It is clearly impossible to measure pressure and temperature in subduction zones directly. But the distribution of temperatures within the lithosphere can be calculated from the heat flow measured on the surface (§1). As for the pressures, they are estimated from the depth by assuming that they result from the burden of the overlying rocks.

On the other hand, on the basis of experimental work, petrologists have proposed stability regions as a function of temperature and pressure for these 'metamorphic facies' (Fig. 5.11). Having a 'model' of pressure and temperature distribution in the

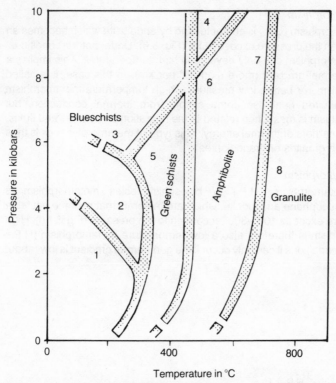

Fig. 5.11 Distribution of metamorphic facies according to pressure and temperature. 1: zeolite; 2: prehnite-pumpellyite; 3: blueschists; 4: eclogite; 5: greenschists; 6 and 7: amphibolite; 8: granulite. (After Ernst, 1974.)

subduction zones, it is therefore possible to locate areas where metamorphic features could occur in both the over-riding and descending plates, and to forecast the type of metamorphism generated (Fig. 5.12). Of course the validity of such an undertaking depends entirely on the models used. Hypotheses take up a large part of all such models and, in particular, it is always difficult to explain natural phenomena through the results of experimental petrology.

The site of high pressure–low temperature metamorphism (blueschist facies)

High pressures are necessary before the jadeite-quartz association appears at the expense of albite (Fig. 5.10). On the models, the site of the H.P.–L.T. metamorphism is usually placed in the subduction zone about 15 km or more deep. The rocks affected by this type of metamorphism belong to the oceanic crust in the descending plate and to the deepest part of the tectonic accretionary prism (deformed turbidites and ocean sediments variously enriched by ophiolite wedges). In contrast, in the over-riding plate the temperature is too high in principle for the H.P. minerals to develop.

Thus considerable tectonic movements must take place after metamorphism for the blueschists to be exposed at the surface:
 (i) *Vertical movements* caused by a return to isostatic equilibrium when subduction slows down or halts (Ch. 4). It is however improbable that an uplift of 15 km could be caused solely by this phenomenon.
 (ii) *Tangential and vertical movements* caused by a collision (Ch. 4).
   In normal conditions, the blueschists would not be accessible for observation on

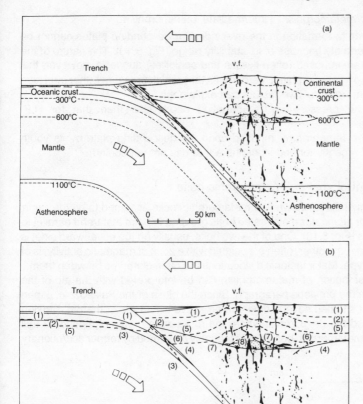

Fig. 5.12 A model illustrating a possible distribution of metamorphic facies in a lithospheric convergence zone. (a) thermal structure of the lithospheric convergence zone; (b) spatial distribution of metamorphic facies in the lithospheric convergence zone; 1: zeolite; 2: prehnite-pumpellyite; 3: blueschists; 4: eclogite; 5: greenschists; 6 and 7: amphibolite; 8: granulite. (After Ernst, 1974.)

active island arcs and margins. This discussion will be resumed in Chapter 6. But it should be noted at once that the pressures caused by tectonic stresses are not considered in the model in Fig. 5.12. Moreover, if such stresses play any part in H.P. metamorphism, then it could be initiated less deeply.

The site of high temperature metamorphism (greenschist and amphibolite facies)

The high temperatures required for the development of this metamorphic type occur in the crust of the volcanic arc about 10 km deep or more (Fig. 5.12). As with high pressure metamorphism, therefore, great vertical movements causing considerable erosion of the crustal surface are needed before these metamorphic rocks can be exposed. Erosion on such a scale (10 km or more) is hard to explain solely by the uplift of the volcanic arc during subduction. In this case again, lithospheric convergence can account for the regional metamorphism, but cannot explain the exposure of the metamorphic rocks. This exposure must take place at a subsequent stage of arc and margin evolution during and after their collision (Ch. 6).

The site of kyanite metamorphism ('intermediate' metamorphism)

The location of kyanite formation in the over-riding or descending plates cannot be determined with certainty because of its stablility range (Fig. 5.10). The nature of the rocks that are metamorphosed (often flyschs and ophiolites) suggests, however, that the site of this type of metamorphism merges with that of high pressure rather than high temperature metamorphism. Two factors can explain this:

(i) *A relatively low pressure* i.e. a formation depth less than that for H.P. metamorphism.

(ii) *A relatively high temperature* imposed upon the descending plate by its youth (subduction of recently formed lithosphere that has not yet cooled).

(c) Regional control: 'paired' metamorphic belts

Japanese petrologists have shown that metamorphic rocks assumed to be the *same age*, but of *different facies*, form elongated belts grouped in pairs that trend parallel to the main structures (Fig. 5.13). One of the two belts, usually on the oceanward side, is of high pressure type; the other, often associated with a zone of magmatic activity, is of high temperature type. Major tectonic dislocations are often observed between them.

This special distribution of metamorphism can be interpreted with the aid of the model described in the previous paragraph. Since the close of the Palaeozoic, Japan has probably been situated several times on an active island arc or margin. Erosion could have exposed the deep zones where the rocks were metamorphosed. In this case, the 'pairs' of metamorphic belts would probably be formed by former accretionary

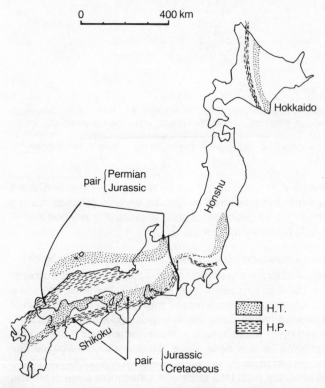

Fig. 5.13 Three paired metamorphic belts in Japan. (After Miyashiro, 1972.)

prism (H.P.)–volcanic arc (H.T.) groups, which would thus reveal the old directions of subduction.

However this interpretation runs into several difficulties:

• Even in Japan it is not certain that the H.P. and H.T. belts, when both exist, are of the same age.

• Most frequently the high pressure belts have no parallel equivalents in the H.T. facies.

• Lastly, the volcanic arc is an area which is being uplifted as long as subduction lasts. But high temperature metamorphism affects sedimentary sequences that are often very thick in the fold mountains, which implies active subsidence.

This last objection, however, is only valid if it is assumed that all thermal metamorphism results from lithospheric convergence, which is certainly erroneous.

In summary, H.P.–L.T. metamorphism probably develops *in the subduction zone* and, in principle, at a depth of about 15 kilometres or a little less. It affects *the deformed sediments in the accretionary prism and the oceanic crust* (ophiolites) *of the descending plate*.

This type of metamorphism is certainly associated with intense deformation, (pressure plays a paramount role), which can go so far as to form 'tectonic mélanges' where the sediments and the fragments of the ocean crust no longer display any organised structure.

Moreover, the heat flow and magmatism that characterise the active arcs and margins strongly suggest that H.T.–L.P. metamorphism takes place *beneath the volcanic arc*.

But this attractive model should not be generalised to such a degree that every regional metamorphism is interpreted as the effect of old subductions. It is highly likely that other processes, and especially those associated with collision (Ch. 6), could have similar metamorphic effects.

Notes

1. 1 H.F.U. $= 1 \ \mu \text{cal cm}^{-2} \text{s}^{-1}$

Further reading

Cawthorn R. C., 1977. Petrological aspects of the correlation between potash content of orogenic magmas and earthquake depth. *Mineral. Mag.,* **41**, pp. 173–182.

Ernst W. G., 1974. Metamorphic and ancient continental margins, in *The Geology of Continental Margins*. Burke C. A. and Drake C. L. (eds). Springer-Verlag, New York, pp. 907–919.

Miyashiro A., 1972. Metamorphism and related magmatism in plate tectonics. *Amer. J. Sci.,* **272**, pp. 629–656.

Miyashiro A., 1974. Volcanic rock series in island arcs and active continental margins. *Amer. J. Sci.,* **274**, pp. 321–355.

Ringwood A. E., 1974. The petrological evolution of island arc systems. *J. Geol. Soc. London,* **130**, pp 183–204.

Ringwood A. E., 1977. Petrogenesis in island arc systems, in *Island arcs, deep sea trenches and back-arc basins.* Talwani M. and Pitman W. C. III. (eds) *Maurice Ewing Series 1*, Amer. Geophys. Union, pp. 311–324.

Chapter 6
Collision and the formation of fold mountains. Continental margins and geosynclines

The first part of this last chapter describes the effects of collision between continents and/or island arcs (§1): the emplacement of basement nappes, including ophiolite nappes; metamorphism; the indentation and crushing of continental margins or island arcs; uplift (orogenesis) and subsidence (in the molasse basins) after the paroxysmal tectonic phase.

In the second part (§2), the old geosynclinal model is compared with the model suggested by the new plate theory. In spite of apparent divergences, both models in fact describe almost the same things. This close correspondence opens the way to palaeo-oceanographic studies (§3).

1 Collision and associated phenomena

Subduction is determined essentially by the contrast in density between the lithosphere and the asthenosphere (Chs. 1 and 4). When the average density of a plate is higher than that of the asthenosphere (as with an oceanic plate), then its descent is possible. In the opposite case (as with a continental plate with a thick and relatively light crust), the lithosphere generally remains on the earth's surface.

The majority of plates, however, have both oceanic and continental parts (Ch. 1). The inevitable consequence of their convergence therefore is that continental crust belonging to the descending plate is transported, at some time or other in geological history, up to the ocean trench. In principle this should stop the subduction and cause a *collision*.

(a) The emplacement of basement nappes

In reality, the convergence of two plates does not stop at once when, for example, a stable margin comes into contact with an active island arc or margin. Thus, for a brief period before subduction stops completely, continental crust may be inserted beneath the over-riding plate. Conversely, collision in theory has the effect of thrusting the leading edge of the over-riding plate on to the other plate and of emplacing thick *basement nappes* upon a continental crust.

According to this outline, which is accepted by many geologists and geophysicists, the nature of the thrust basement directly depends on the nature of the lithosphere on the leading edge of the over-riding plate. If it has oceanic characteristics, then the collision causes the emplacement of ophiolite nappes[1]. On the other hand, when the leading edge of the over-riding plate has a sialic crust, then the same phenomenon causes the formation of *continental basement nappes*.

These basement nappes are basically composed of crustal material. But it sometimes happens that mantle rocks (peridotites) also take part in the thrusts. It is, however, obvious that only the upper part of the lithosphere is involved. This suggests that the upper part could be decoupled from the lower part when collision occurs. A decoupling horizon, whose existence is suggested by seismic studies, should thus be sought within the mantle itself. It will be recalled that the lithosphere seems to be divided into two strata by a 'low velocity channel' which is interpreted as a zone of serpentinised peridotites by some geophysicists. Serpentinites are very plastic rocks and if it were to be proved that they do in fact form a continuous layer, they would certainly facilitate the decoupling of the upper lithosphere (crust + mantle sole) from the lower lithosphere. In such conditions, the tangential forces acting during a collision could, in principle, give rise to basement nappes composed of both crustal material and mantle rocks.

In theory, three types of nappe can be found:

Ophiolite nappes

This case is illustrated by the model proposed in Fig. 6.1. Before the collision, the substratum of the area between the trench and the volcanic arc is assumed to be oceanic in nature. When the continental crust on the descending plate is dragged into the subduction zone, the oceanic crust and its mantle sole bordering the over-riding plate can give rise to nappes of ophiolites of considerable thickness. The great ophiolite overthrusts described in the fold mountains and on certain Pacific islands could be explained in this way. In this interpretation it will be seen that the thrusting occurs just before the convergence stops completely. It is almost exactly contemporaneous with the collision and the tectonic paroxysm.

Nappes of intermediate crust

The change from a stable to an active margin (*cf.* 6, p. 53) probably has the effect of leaving the 'intermediate' crust at the front of the over-riding plate. This 'intermediate' crust formerly constituted the transition between continent and ocean (*cf.* 6, p. 31). During subduction, this particular crust (that was inherited from the previous history of the margin), occupies an identical position to that of the oceanic crust placed on the edge of the over-riding plate in Figs 6.1 and 6.18. If collision occurs, the basement of the

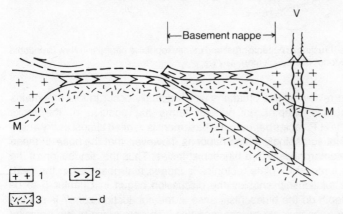

Fig. 6.1 A model for the emplacement of ophiolite nappes. 1: continental crust; 2: oceanic crust; 3: subcrustal mantle sole; d: decoupling zone (serpentinite layer); M: Moho; V: volcanic arc. The basement nappe (here ophiolitic in character) has not yet been emplaced at this stage.

old stable margin and the peridotites composing its base can theoretically take part in crustal thrusting which is wholly comparable to the thrusting of the oceanic crust and mantle. The problem of the nature of the intermediate crust arises once more (Ch. 2), and it will thus be understood that the correct interpretation of the great basement thrusts observed in the fold mountains depends on which solution is adopted. Are the characteristics of the intermediate crust more oceanic than continental? Then some ophiolite nappes may perhaps belong to it. Or is this intermediate crust composed of deep continental crust brought close to the surface by tensional phenomena (Fig. 2.16)? In this case, the granulite basement nappes should be related to the intermediate crust. These nappes are often associated with ultramafic rocks described in many fold mountains (Ivrea zone in the western Alps, Calabrian area in southern Italy, Betic and Rif chains, Pyrenees, etc.).

The thrusts of typical continental crust lastly, can also occur if the descending plate is inserted beneath a thick sialic crust, as, for example, along a Cordillera-type active margin.

(b) Distribution of metamorphism

The over-riding plate should only undergo an H.T. metamorphism during subduction (Ch. 5). The basement nappes belong to that plate when they are emplaced and cannot, in principle undergo H.P.–L.T. metamorphism unless it be restricted to their very base. Strong stresses at low temperatures, on the other hand, act on the deep part of the tectonic accretionary prism and upon the descending plate inserted beneath the basement nappes. In geological language, it is therefore only the relative autochthon (the continental crust on the descending plate) and the sedimentary sole of the nappes (the laminated remains of the accretionary prism) which ought to contain the blueschists.

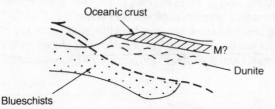

Fig. 6.2 Diagram of the thrusting of oceanic crust and mantle (ophiolite nappes) in New Caledonia and New Guinea. M: Moho. (After Coleman, 1971.)

It is impossible to review all the situations which actually occur in fold mountains within the limits of this chapter. In general terms the position of the rocks metamorphosed in the H.P. facies *beneath* the basement is verified almost everywhere (Figs 6.2 and 6.3). Not surprisingly it also happens, however, that the *base* of these nappes displays metamorphites with a blueschist facies. Thus the distribution of the metamorphic facies forecast by plate tectonics is indeed that observed in the field. Moreover this good relationship enables the discussion begun in Chapter 5 to be resumed: At what depth do the blueschists arise in the subduction zone? Is not the 15 km suggested by geophysicists an exaggeration? The thickness of the ophiolite nappes cannot exceed the thickness of the oceanic crust (7 km) and its mantle sole (a few kilometres at most), i.e. a maximum of about 12 kilometres. The blueschist facies

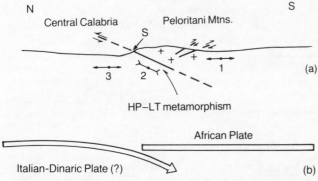

Fig. 6.3 The Calabrian suture. (a) schematic cross-section; (b) interpretation. Here the basement nappe is composed of continental material. S: suture; 1: location of Jurassic extension (basalts) (back-arc basin?); 2: location of Cretaceous-Eocene compression and H.P.–L.T. metamorphism; 3: location of end-Cretaceous extension (alkaline basalts) (effects of the bending in the descending plate?). (After Dubois, 1976.)

appears beneath the nappes in very many areas, every time that post-tectonic erosion enables the autochthonous rocks to be studied. It seems therefore that high pressure metamorphism can appear in subduction zones about 10 kilometres deep, or even closer to the surface. It seems also that the amount of crustal erosion required to expose the blueschists is less than that envisaged by present geophysical models.

(c) 'Indentation' and the crushing of continental margins

When two plates carrying continental crust converge, their collision must start in those areas where the continents jut out into a salient (an 'indenter'). Meanwhile subduction, in contrast, continues where oceanic crust still remains between the approaching margins. Thus the plate boundary consists of some segments that are subject to considerable tangential forces (in collision zones) whilst other segments offer but little resistance to convergence (in the subduction zones).

This situation could have occurred many times during the course of geological history. It has recently been the subject of a theoretical study which opens the way to a coherent interpretation of lithospheric sutures[2] (Fig. 6.4). The proposed model is based on the analysis of the deformation of 'soft' metals. Their behaviour is compared to that of the lithosphere in the period between the start of the collision and the point when plate convergence comes to a complete halt. Two examples of its possible applications are suggested in Fig. 6.4:

Continental indentation

A linear active margin is telescoped by a spur of continental crust. The active margin is intensely deformed where the collision occurs (Fig. 6.4(a)).

Continental crushing

A linear stable margin plunges beneath an active margin with a salient. This salient is crushed up against the stable margin (Fig. 6.4(b)).

It will be noted that it is always the over-riding plate that undergoes the most intense deformation when collision occurs[3]. In both examples quoted the intracontinental overthrusts appear on this plate, whilst new subduction zones form secondary arcs laterally. When the lithospheric convergence finally ceases, the suture zone in the end

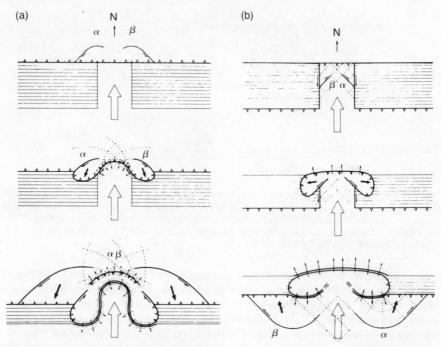

Fig. 6.4 Two types of plane deformation associated with continental collision. Subduction zones (open chevrons), intracontinental overthrusts (closed chevrons) and transcurrent faults are represented by thick lines; folds in dotted lines; lines of sliding in dashed lines (α: dextral; β: sinistral); the small arrows represent the direction of pitching; horizontal lines represent the oceanic crust; and the bold double line represents the suture zones. (After Tapponier, 1977.)

provides a very deformed picture of the geometry of the continents before they collided.

These models provide a satisfactory interpretation of certain distinctive features of the Alpine chain in Europe and Asia, which runs in a succession of arcs and festoons that probably appeared when Africa, Arabia and India collided with the Eurasian plate.

(d) Vertical movements following from collision

From the diagram of collision (Fig. 6.5), the zone where the basement nappes are superimposed upon a thick continental crust must be an area of rapid uplift (*orogenesis*, following from tectogenesis). The preservation of isostatic equilibrium where the crust is considerably thickened implies the development of a *root* (the depression of the Moho), and of great relief features (mountain chains). These mountains are immediately attacked by erosion, which brings about a new uplift by compensating isostatic re-adjustment. The process continues, in principle, as long as the crust has not resumed its normal thickness (30–40 km). The young mountain chain is thus bordered by thick sedimentary accumulations (molasse) which result from its own gradual destruction. Meanwhile the continental crust on the over-riding plate reveals exposures of progressively deeper structural horizons. These can go down as far as the high grade metamorphic rocks forming the base of the crust and even as far as the mantle. Thus, for the second time in this chapter, the granulitic and ultramafic basement appears as an 'index-marker' of collision mountain chains. In the first case (the thrusting of intermediate crust) its exposure must pre-date the geosynclinal sedimentation and the thinned continental crust would then form part of the flysch

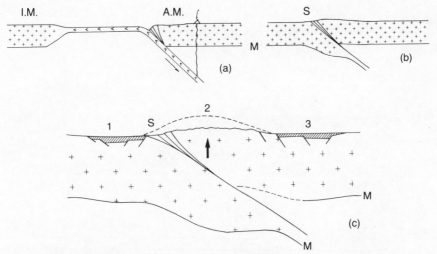

Fig. 6.5 Schematic diagram of continent– continent collision. (a) subduction before the collision; (b) and (c) diagram of the collision chain; 1: external molasse trough; 2: mountain chain; 3: internal molasse trough; M: moho; R: root; I.M.: inactive margin: A.M.: active margin; S: suture.

basement. In the second case, the rocks formed at the base of the sialic crust would only be eventually exposed at the end of the collision after considerable erosion.

The molasse is evacuated to two fault-troughs aligned parallel to the trend of the chain. The external molasse trough is situated on the old descending plate. Conversely the internal molasse trough is aligned along the over-riding plate beyond the zones of crustal superimposition.

The normal faulting on the edges of these troughs may be partly explained by opposing movements on the continental basements. These cause uplift in the mountain chain and sinking beneath the weight of sediments in the molasse basins. It is probable, however, that the effects of active crustal extension should be added to those of isostatic readjustment. The external molasse trough, especially, can to some degree be likened to the fault-troughs which appear on the oceanward side of a marginal trench, where the descending plate bends before plunging beneath the over-riding plate (*cf.* 1, p. 58).

(e) Mountain chains formed by collision

Chains formed by collision occur in different situations according to whether the suture zones bring former continental margins or remnant island arcs into contact:

 • The collision of two island arcs gives rise to an *intra-oceanic fold mountain chain* (Fig. 6.6). A fine example is provided by the Molucca Sea in Indonesia (Talang-Mayu ridge).

 • The collision between an island arc and a continental margin creates a chain *situated on a continental border*. This is in a similar position to a subduction chain but its geological characteristics are markedly different (Fig. 6.7). The Californian chain was probably formed in this fashion.

 • Lastly, the collision between two margins (Fig. 6.5) gives rise to *continent-continent collision chains*, which occupy the suture line between two plates that henceforth are welded together. This is the case in the majority of the mountain chains that cross the continents, such as the Himalayas, the Alps and even the

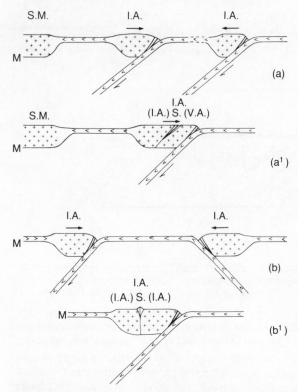

Fig. 6.6 Schematic diagram of island arc–island arc collision. I.A.: island arc; S.M.: stable margin; S: suture; M: Moho; V.A.: Volcanic Arc.

Pyrenees. In general, one of the margins was active before the collision, whilst the other was stable. This situation was described at the beginning of this chapter in paragraphs (a), (b), (c) and (d). But a collision between two active margins is also possible in theory.

2 Continental margins and geosynclines

Most fold mountains are thus composed of former continental margins which have been deformed and have emerged as a result of plate collision. In principle therefore, the subject of this book extends far beyond the domain of the present oceans. It is hard, however, to introduce the fold mountains in a few pages. On the other hand, another approach is possible. For several decades geologists have compared their regional syntheses with the geosynclinal 'model' which accounted for most of their observations. This model appears in fact to be far removed from that presented by plate tectonics. It is therefore important to compare it with the 'actualistic' model, i.e. with the continental margins as they have been introduced in the preceding chapters.

(a) Summary of the geosynclinal 'model' (Fig. 6.8)

A geosyncline in principle is composed of two 'troughs' and two 'ridges'. Going from the continental dry land (the foreland) towards the ocean, the following zones are encountered in succession:

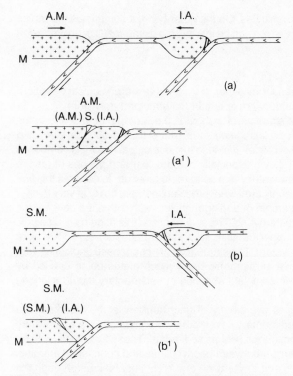

Fig. 6.7 Two examples of island arc–active continental margin collision. A.M.: active margin; I.A.: island arc; S: suture; M: Moho; S.M.: stable margin.

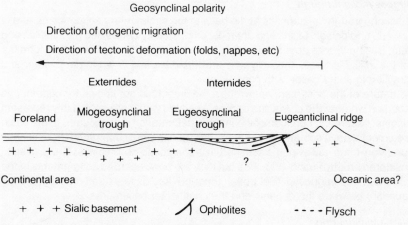

Fig. 6.8 Model of a geosyncline. (After Aubouin, 1965.)

The 'miogeosynclinal trough'

This is an elongated basin about 1 000 km long and 200 km wide. Thick sedimentary sequences accumulate within it composed of *neritic or hemipelagic facies* which may exceed several kilometres in thickness. It is only at the end of the evolution, after the tectogenesis of the internal zones, that the flyschs (turbidites) are deposited above the neritic sequence.

Collision and the formation of fold mountains

This first trough contains no ophiolites. On the other hand, it sometimes preserves traces of some magmatic activity which can be interpreted as the result of an extension of the continental crust.

The miogeanticlinal ridge

In different areas this ridge corresponds either to a shoal (its morphological definition) or to an area where the sedimentary cover is thin (its structural definition).

(i) The most frequently quoted example of a shoal is the *Gavrovo geanticline* in the Dinaric chain. It is a markedly elongated area (more than 1 000 km long) and relatively narrow (100–150 km wide). Its Mesozoic geological history is composed of episodes of emergence (shown especially by bauxites) and periods of marine sedimentation in very shallow water when coral reefs grew up. In general, this first type of geanticline is a *subsiding* area where the sedimentary cover is very thick.

(ii) The most representative example of a miogeanticline with a thinned cover is the *Briançonnais ridge* in the western Alps. The sediments on this ridge have a pelagic facies (at least during the upper Jurassic and the Cretaceous), and they were deposited in relatively deep waters. The sequences are condensed (indicating very slow sedimentation), or have stratigraphic gaps (each hiatus being caused by submarine erosion, or by the complete absence of sedimentary input for a long period).

Miogeanticlines thus belong to two clearly distinct families: some belong to the neritic zone and have a thick sedimentary cover; others are situated in deeper waters and only receive thin and discontinuous pelagic sediments. These two types must be clearly separated in any comparison between geosynclines and continental margins. On the other hand, they both belong to the zone of the continental crust through the nature of their basements.

The eugeosynclinal trough

In this second trough the deposits at the base of the sedimentary sequence generally have a deep and pelagic facies and are most often entirely decalcified (radiolarites and argilites). But the flyschs appear very early in the history of the basin and they form the bulk of the infill. Generally these flyschs are turbidites laid down in deep water and interstratified in some cases with contourites.

The nature of the basement below these sediments is not known for certain. It is never certain whether the crystalline rocks associated with the flyschs really represent their original basement, because of the intense deformation that occurred in the basin after sedimentation. It sometimes happens that the apparent substratum is continental in nature. Very often however, the substratum is composed of ophiolites. These used to be considered eruptive rocks (Fig. 6.8), but they are now regarded as fragments of the oceanic crust. The eugeosynclinal trough therefore seems to extend on either side of the boundary between the oceanic and the continental basements.

The eugeanticlinal ridge

The presence of this ridge is often suggested by the direction and the nature of the sedimentary input in the eugeosyncline, which implies a source area of continental crust situated in an 'internal' position in relation to the basin.

The internides

These are composed of the eugeanticline–eugeosyncline group. These are the first zones, from the point of view of geological history, to undergo compressional deformation. Very early movements often occur on the eugeanticlinal ridge or on the

102

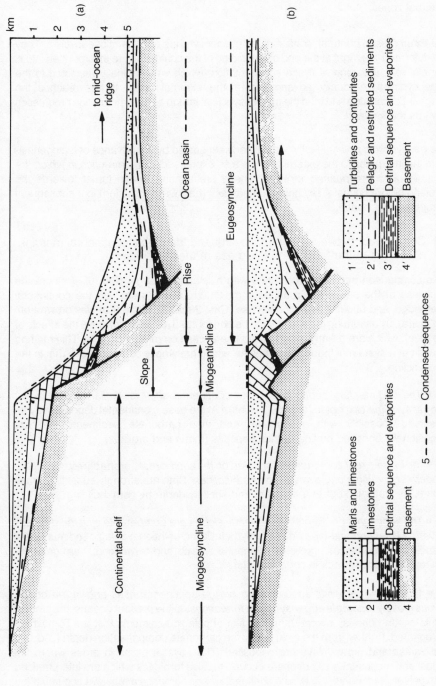

Fig. 6.9 Comparison between the geosynclinal model and a stable continental margin. (a) outline of a 'starved' margin; (b) same outline drawn as if the sea floor were horizontal. The different palaeogeographic units in the geosyncline are clearly brought about.

internal edge of the eugeosyncline (*'palaeotectonic' movements*). Subsequently, when the mountain chain itself is upheaved (in the *tectonic paroxysm*), the *'wave of tectogenesis'* starts from the most internal zones and then migrates towards the external zones.

The externides in contrast, consist of the miogeosyncline and the miogeanticline. They are deformed late, right at the end of the folding of the chain, and are incorporated into it. Correlatively, the zone of flysch sedimentation, which was originally restricted to the eugeosyncline, gradually advances over the external zone. It first reaches the miogeanticlinal ridge and then the miogeosynclinal trough as the deformation progresses towards the foreland.

The geosynclinal polarity lastly, was already suggested by the advance of tectogenesis from the internides to the externides. It is also expressed by *the direction in which the tectonic units are pushed forward*. These are almost always thrust towards the foreland before occasionally being turned back (backward-thrusting) in a subsequent compressional phase.

(b) Comparison between the geosyncline and the stable continental margins according to the facies and thickness of their sediments

This comparison can easily be made with a 'starved' margin (Fig. 6.9), where the thickness of the sediments reaches a maximum in two areas: on the continental shelf-edge, and under the continental rise (Chs. 2 and 3). Fig. 6.9(b) was drawn from Fig. 6.9(a), by assuming the sea-floor to be horizontal. This was to remove the effects of the relief, which are greatly exaggerated in any case on most diagrams. This method brings out a system of 'ridges' and 'basins' which are wholly comparable to that in the geosyncline.

• *The sedimentary sequence on the continental shelf is similar to that in the miogeosyncline* (except for the late flyschs).[4] At the base, continental deposits may be observed (possibly with evaporites), and then carbonate sediments that are occasionally enriched by fine detrital deposits (marls and argilites).

• *The barrier-reef zones usually situated on the shelf-break, probably represent the modern equivalents of Gavrovo-type geanticlines.* Both are strongly subsiding areas where the water is kept at a constant and shallow depth by coral building.

• *The sediments on 'starved' continental slopes are comparable with the deposits covering Briançonnais-type ridges* (polymetallic incrustations; pelagic sediments and 'condensed' sequences; traces of submarine erosion and re-working; 'hard grounds', where the exposed rock is corroded, etc).

• Finally, *the deposits accumulating on the continental rises and in the ocean basins closely resemble eugeosynclinal formations*. In the present oceans, the crest of a mid-ocean ridge is covered by a thin film of pelagic sediments that are sometimes carbonated. Further from the axis, below the carbonate compensation depth (C.C.D.), radiolarites and brown clays are deposited. The abyssal plains in areas where the oldest and most sunken lithosphere occurs, receive turbidites which are interstratified with pelagic sediments. Lastly, the continental rises comprise a massive accumulation of turbidites and contourites which are probably the equivalents of the flyschs.

(*c*) Comparison between the geosyncline and the continental margins according to the nature of their basement

Like the basement of the continental shelf and slope, the basement of the miogeosyncline and the miogeanticline is *continental* in nature. In different areas, the eugeosyncline extends over either a *continental or an oceanic basement*, in the same way as the continental rise is built on the ocean–continent transition zone. The comparison between the basements of stable margins and those of geosynclines therefore suggests the same relationships as the facies and thickness of the sediments. It remains only to identify the *eugeanticlinal ridge*. This ridge, in the geosynclinal model, constitutes an area with a *continental substratum*, which provides part of the detrital supply to the flyschs. It cannot therefore be the mid-ocean ridge which has a basaltic basement. On the other hand, its characteristics clearly link it to the *island arcs*. They both have the same continental basement; both are situated near the ocean floors (between a descending oceanic plate and a marginal basin); and lastly, both have the same general configuration – an elongated ridge with islands that are being upheaved and eroded.

(*d*) Comparison between the geosyncline and the active island arcs and margins

If the eugeanticlinal ridge is indeed the equivalent of an island arc, where then should the eugeosyncline be placed in the actualistic comparative outline?

In the marginal basin? The external zones would then coincide with the edge of the basin situated opposite the volcanic arc.

In the ocean carried by the descending plate? In this case the miogeosynclinal and miogeanticlinal zones (the externides) would correspond with the stable margin of this ocean, which would move towards the eugeanticlinal ridge until collision occurred. Thus the eugeosyncline would not only be linked to the remnants of the disappearing ocean, but also to the sediments and ophiolites in the tectonic accretionary prism that probably come largely from this ocean.

In reality the two varieties of eugeosyncline (the marginal basin and the ocean being shortened by subduction) could have coexisted in the geological past just as they do in nature at present. But they probably correspond to different times in geosynclinal history. The marginal basin rather represents a youthful stage which occurs well before the formation of a fold mountain chain. In contrast the shrinking of an ocean that belongs to a plate descending beneath a volcanic arc comes only shortly before that arc collides with the margin opposite it. For example, the upper Jurassic tectogenesis affecting the internal zones of the Jugoslavian Dinarides (Fig. 6.10), could have been caused by the subduction of an ocean area beneath an ophiolite-rich tectonic accretionary prism and beneath the volcanic arc of the Rhodope margin. It will be noted from this diagram that the external areas (the Apulian stable margin for example) are not deformed until collision occurs.

Thus *old tectonic movements*, affecting the most internal units in the geosyncline, can be interpreted without difficulty in the plate tectonic model by the subduction of the geosynclinal area beneath the eugeanticlinal ridge (or island arc). This model, moreover, explains *the gradual advance of tectogenesis towards the external zones* and that *the tectonic units point towards the foreland*.

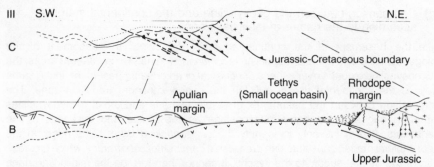

Fig. 6.10 Two stages of old ocean development in part of the Jugoslavian Dinarides. (After Blanchet, 1976.)

(e) Collision and the evolution of the geosyncline

According to the model suggested by plate theory, two tectonic events should, in principle, occur in succession as a result of a continent–continent collision:

Deformation of both continental margins through intense compression with at least one margin being active. This is the *paroxysmal tectonic phase*. In spite of its deformation however, the folded chain preserves the imprint of its composite origin (two margins crushed and welded together) in its structural arrangement. It displays a certain symmetry in relation to the suture zone that geologists had noted before they could interpret it with the aid of the actualistic model. In this respect, the vocabulary they used to describe orogens is most revealing. Chains were described as 'bimarginal', 'bilateral', 'back-to-back', etc.

Vertical movements determined by isostatic readjustments (creation of molasse troughs on either side of the mountain chain). This is the *late tectonic* event described by structural geologists. At this stage the tilting of crustal blocks that have undergone differential uplift may cause the *cover to slide*. This especially involves the external zones and chiefly the miogeosynclinal sediments. These have been less affected by the paroxysmal tectonic phase than the internal zones because of their distal position in relation to the suture zone.

Thus plate tectonics not only enables the sedimentary history of the geosyncline to be re-interpreted (*cf.* (*b*) p. 104) but it also gives just as satisfactory an explanation of the tectonic history and structural arrangement of the fold mountains. Table 2 summarises this discussion and illustrates how the geosynclinal system, as understood by geologists, can be compared with the arrangement of the present oceans (allowing for a few morphological dissimilarities). There is therefore no reason to set the geosynclinal system in opposition to the model suggested by plate theory, as both describe the same objects with different words. The picture of the geosyncline should, however, be 'touched up' in two ways to bring it into line with the outline of the present oceans and continental margins: (i) the ophiolites are most often relics of the floors of the major oceans or marginal basins; (ii) the Briançonnais-type miogeanticline is not a ridge in the morphological sense of the term, but it represents rather a former continental slope with a thin sedimentary cover.

3 Conclusion: subduction chains and collision chains

(*a*) If a mountain chain is defined by the existence of *relief*, then clearly subduction,

Table 2   The geosynclinal system and the present oceans

Nature of the crust (basement of the sediments)	Geosyncline	Equivalent in the present oceans.
Thick continental crust	Foreland	Emerged continent
	Miogeosyncline	Continental shelf
	Gavrovo-type Miogeanticline	Continental shelf-edge (barrier reefs)
	Briançonnais-type Miogeanticline	Continental slope
Thinned continental Crust / Oceanic crust	Eugeosyncline	Continental rise / Ocean basin (+ tectonic accretionary prism) or Marginal basin
Continental or intermediate crust	Eugeanticline	Island arc

acting alone, can give rise to an orogen. The Andean Cordillera bears witness to this. Here tangential tectonic movements are slight, except, perhaps, in the tectonic accretionary prism, when it exists. The uplift of the crust and the growth of a root beneath the mountains are probably the results of magmatism caused by subduction. (Hence the name 'thermal chain' that is sometimes given to this kind of orogen.)

(b) Many fold mountains (geosynclinal chains in the true sense) result from *the collision of continental margins or island arcs*. Geologists and geophysicists have described regional examples of the three groups of collision chains previously listed (*cf.* (e) p. 99). But the vast work involved in unravelling the sedimentary and structural history of orogens by following the new guidance offered from global tectonics is far from being completed. The undertaking is all the more delicate because two other categories of orogen exist (apart from subduction chains and collision chains) which have not been considered in this book. These are (i) *strike-slip chains*, whose evolution is determined when two plates slide past each other and bring continental crust into contact (intracontinental transform faults), and (ii) *crushed continental rifts*.

Clearly, those geologists and geophysicists who are interested in continental margins have a field of study which extends well beyond the present oceans and as far as the great continental surfaces occupied by the successions of chains formed in Alpine, Hercynian, Caledonian and even earlier times. Not long ago a few of them left the traditional confines of their study and embarked on boats to look for the continuations of the dry land beneath the sea. What they found, instead of what they were expecting, has changed the discipline's whole way of thinking. Henceforth the Earth Sciences are guided by a new theory that was born in the ocean. Marine geophysicists and geologists now join their colleagues on the dry land to look for the traces of lost oceans on the continents. This 'palaeo-oceanography' will undoubtedly be the work of the next decade.

Notes

1. The over-thrusting of oceanic crust has often been designated by the term *'obduction'*. According to its initial definition, obduction is a consequence of subduction. It would thus be the thrusting of oceanic crust and mantle over an active island arc or margin, quite outside any collision phenomenon. But it seems that the obduction so defined does not occur on the earth's surface at present and it would be probably better to abandon the term for that reason.

2. The *suture* between two plates corresponds to their former boundary which is now inactive. It is marked by completely crushed mylonites and tectonic mélanges in which blueschist facies metamorphites may be recognised as well as lenses of ophiolites that are perhaps inherited from the tectonic accretionary prism.

3. Deformation does not, however, spare the descending plate. In the most frequent cases the collision occurs between an active margin and a stable margin (Fig. 6.5) whose former structures can then be re-activated. It should be expected, in particular, that former normal faults should be re-activated as reverse faults during the collision. The tectonic wedges thrust towards the foreland would then correspond to the crustal blocks that sank and tilted during the previous rifting on the margin which took place several tens of millions of years earlier. This interpretation would explain the frequent coincidence between the palaeogeographic units determined by initial tensional movements and the tectonic units which appeared through the influence of subsequent compression.

4. This exception does not constitute an objection: the flyschs occupy the external zone when the internal zones are deformed, i.e. at a stage of evolution that the Atlantic-type stable margins have not yet reached.

Further reading

Aubouin J., 1965. *Geosynclines*. Elsevier Publishing Company, Amsterdam, 335 pp.

Dickinson W. R., 1974. 'Plate tectonics and sedimentation', in Dickinson W. R. (ed.) *Tectonics and Sedimentation,* Soc. Econ. Paleont. Miner., Special Publication, No. 22.

Mattauer M., 1973. *Les déformations de l'écorce terrestre*. Hermann (ed.), Paris, 493 pp.

'Histoire structurale des bassins méditerranéens. International Symposium, Split, Yugoslavia, 25–29 October, 1976. Technip, Paris, 1977, 448 pp.

References

This list is restricted to the works to which reference is made in the figures. In addition, the reader will find other references at the end of each chapter which are designed as a guide to further reading.

Aubouin J., 1965. 'Geosynclines', *Development in Geotectonics*, **1**, Elsevier Publishing Company, Amsterdam, 335 pp.

Barazangi M. and **Isacks B.**, 1971. 'Lateral variations of seismic-waves attenuation in the upper mantle above the inclined earthquake zone of the Tonga Island arc: deep anomaly in the upper mantle', *J. Geophys. Res.*, **76**, pp. 8493–516.

Barberi F., 1974. 'Volcanisme et tectonique des plaques', in *Volcanisme et phénomènes associés*, Journées d'Aurillac, 15 et 16 mai 1974. *Rev. Haute Auvergne*, **44**, pp. 245–76.

Blanchet R., 1976. 'Bassins marginaux et Téthys alpine: de la marge continentale au domaine océanique dans les Dinarides', in *Histoire structurale des bassins méditerranéens*. International Symposium, Split, Yugoslavia, 25–29 October 1976. Technip, Paris, pp. 47–72.

Boillot G., Dupeuble P. A., Hennequin-Marchand I., Lamboy M., Lepretre J. P. and **Musellec P.**, 1974. 'Le rôle des décrochements "tardi-hercyniens" dans l'évolution structurale de la marge continentale et dans la localisation des grands canyons sous-marins à l'Ouest et au Nord de la Péninsule ibérique, *Rev. Géogr. Phys. Géol. Dyn.* (2), **16** (1), pp. 75–86.

Bott M. H. P., 1971. 'Evolution of young continental margins and formation of shelf basins', *Tectonophysics*, **11**, pp. 319–27.

Bourcart J., 1959. 'Morphologie du précontinent, des Pyrénées à la Sardaigne', in *La topographie et la géologie des profondeurs océaniques*. 83e Coll. Internat. du CNRS, Nice-Villefranche, 5–12 mai 1958, pp. 33–52.

Coleman R. G., 1971. 'Plate tectonic emplacement of upper mantle peridotites along continental edges', *J. Geophys. Res.*, **76**, pp.1212–22.

Collot J. Y. and **Misségué F.**, 1977. 'Crustal structure between New Caledonia and the New Hebrides', in *Geodynamics in Southwest Pacific. International Symposium*, Nouméa, New Caledonia, 27 August–2 September 1976. Technip, Paris, pp. 135–43.

Coulomb J. and **Jobert G.**, 1973. *Traité de géophysique interne*. Masson, Paris, 2 vols, 1250 pp.

Dewey J. F. and **Bird J. M.**, 1970. 'Mountain belts and the new global tectonics',*J. Geophys. Res.*, **75**, pp. 2625–47.

Dickinson W. R., 1973. 'Widths of modern arc-trench gaps proportional to past duration of igneous activity in associated magmatic arcs', *J. Geophys. Res.*, **78**, pp. 3376–89.

Dubois R., 1976. 'La suture calabro-appenninique crétacé-éocène et l'ouverture tyrrhénienne neogene: étude pétrographique et structurale de la Calabre centrale'. Thèse de Doctorat ès-Sciences, C.N.R.S. n° 12692, Université Pierre et Marie Curie, Paris, 3 vols, 567 pp., 29 pl.

Ernst W. G., 1974. 'Metamorphism and ancient continental margins', in C. A. Burke and C. L. Drake (eds), *The Geology of Continental Margins*. Springer-Verlag, New York, pp. 907–19.

Estocade (Groupe), 1977. 'Étude par submersible des canyons des Stoechades et de Saint-Tropez', *C. R. Acad. Sci.*, Paris, D, **284**, pp. 1631–4.

Fisher R. L., 1974. 'Pacific-type continental margins', in C. A. Burke and C. L. Drake (eds), *The Geology of Continental Margins*. Springer-Verlag, New York, pp. 25–41.

Gennesseaux M. and **Vanney J. R.**, 1979. Cartes ballymetriques du bassin algero-provençal. *Comptes Rendus somm. soc. geol. Fr.*, **4**, pp. 191–4.

References

Hatherton T., 1974. 'Active continental margins and island arcs', in C. A. Burke and C. L. Drake (eds), *The Geology of Continental Margins*. Springer-Verlag, New York, pp. 93–103.

Hayes D. E., 1966. 'A geophysical interpretation of the Peru-Chile Trench', *Marine Geol.*, **4**, pp. 309–51.

Hayes D. E. and **Ewing M.**, 1970. 'Pacific boundary structure', in *The Sea: ideas and observations on progress in the study of the seas*, A. E. Maxwell (ed.), vol. 4, part 2. Wiley, Interscience, New York, pp. 29–72.

Heezen B. C. and **Tharp M.**, 1961. *Physiographic Diagram of the South Atlantic Ocean*. Geol. Soc. America, New York.

Heezen B. C. and **Tharp M.**, 1964. *Physiographic Diagram of the Indian Ocean*. Geol. Soc. America, New York.

Heezen B. C., Tharp M. and **Ewing M.**, 1959. *The floor of the oceans.1: The North Atlantic*. Text to accompany the physiographic diagram of the North Atlantic. Geol. Soc. America, Special paper, **65**, 122 pp.

Hirlemann G., 1974. 'Block-faulting at a Rhine graben scarp and a comparable recent landslide structure at the korinthian graben', in J. H. Illies and K. Fuchs (eds), *Approaches to Taphrogenesis, Proceedings of an International Rift Symposium*, Karlsruhe, 13–15 April 1972. E. Schweizerbart'sche Verlagsbuchhandlung, Stuttgart, pp. 172–6.

Hirn A., 1976. 'Sondages sismiques profonds en France', *Bull. Soc. Géol. Fr.* (7), **23** (5), pp. 1065–71.

Holmes A., 1965. *Principles of Physical Geology*. Nelson, Edinburgh, 1288 pp.

Illies J. H., 1974. Introductory remarks, in J. H. Illies and K. Fuchs (eds), *Approaches to Taphrogenesis, Proceedings of an International Rift Symposium*, Karlsruhe, 13–15 April 1972. E. Schweizerbart'sche Verlagsbuchhandlung, Stuttgart, pp. 1–13.

Isacks B. L. and **Barazangi M.**, 1977. 'Geometry of Benioff zones: lateral segmentation and downward bending of the subducted lithosphere', in M. Talwani and W. C. Pitman III (eds), *Island arcs deep sea trenches and back-arc basins. Maurice Ewing Symp., Series 1*, Amer. Geophys. Union, Washington D.C., pp. 99–114.

Isacks B. L., Oliver J. and **Sykes L. R.**, 1968. 'Seismology and the new global tectonics', *J. Geophys. Res.*, **73**, pp. 5855–99.

Karig D. E. and **Sharman G. F. III.**, 1975. 'Subduction and accretion in trenches', *Geol. Soc. Am. Bull.*, **86**, pp. 377–89.

LeFevre C., Dupuy C. and **Coulon C.**, 1974. 'Le volcanisme andésitique', in *Volcanisme et phénomènes associés*, Journées d'Aurillac, 15 et 16 mai 1974, *Rev. Haute Auvergne*, **44**, pp. 313–355.

Le Pichon X., 1973. 'Introduction sommaire à la tectonique des plaques', in J. Coulomb and G. Jobert, *Traité de Géophysique Interne, II: Magnétisme et Géodynamique*. Masson, Paris, pp. 403–48.

Le Pichon X., Francheteau J. and **Bonnin J.**, 1973. 'Plate tectonics', *Development in Geotectonics*, **6**. Elsevier Publishing Company, Amsterdam, 300 pp.

Lliboutry L., 1973. 'Isostasie. Propriétés rhéologiques du manteau supérieur', in J. Colomb and G. Jobert, *Traité de géophysique interne, 1, sismologie et pesanteur*. Masson, Paris, pp. 473–506.

Lowell J. D. and **Genik G. J.**, 1972. 'Sea-floor spreading and structural evolution of Southern Red Sea', *Amer. Ass. Petr. Geol. Bull.*, **56**, pp. 247–59.

Miyashiro A., 1972. 'Metamorphism and related magmatism in plate tectonics', *Am. J. Sci.*, **272**, pp. 629–56.

Miyashiro A., 1974. 'Volcanic rock series in island arcs and active continental margins', *Am. J. Sci.*, **274**, pp. 321–55.

Nielson D. R. and **Stoiber R. E.**, 1973. 'Relationship of potassium content in andesitic lavas and depth to the seismic zone', *J. Geophys. Res.*, **78**, pp. 6887–92.

Parga J. R., 1969. 'Spätvariszische Bruchsysteme im Hesperischen Massiv'. *Geol. Rundschau*, **59**, pp. 323–36.

Piper D. J. W., Von Huene R. and **Duncan J. R.**, 1973. 'Late quaternary sedimentation in the active eastern Aleutian trench', *Geology*, **1**, pp. 19–22.

Press F., 1970. 'Regionalized Earth models', *J. Geophys. Res.*, **75**, pp. 6575–81.

Rabinowitz P. D., 1974. 'The boundary between oceanic and continental crust in the Western-North Atlantic', in C. A. Burke and C. L. Drake (eds), *The Geology of Continental Margins*. Springer-Verlag, New York, pp. 67–84.

Ringwood A. E., 1974. 'The petrological evolution of island arc systems', *J. Geol. Soc. London*, **130**, pp. 183–204.

Sclater J. G., Anderson R. N. and **Bell M. L.**, 1971. 'Elevation of ridges and evolution of the central eastern Pacific', *J. Geophys. Res.*, **76**, pp. 7888–915.

Seely D. R., Vail P. R. and **Walton G. G.**, 1974. 'Trench slopes model', in C. A. Burke and C. L. Drake (eds), *The Geology of Continental Margins*. Springer-Verlag, New York, pp. 249–60.
Sheridan R. E., 1976. 'Sedimentary basins of the Atlantic margin of North America', *Tectonophysics*, **36**, pp. 113–32.
Sleep N. H., 1971. 'Thermal effect of the formation of Atlantic continental margins by continental break up', *Geophys. J. Roy. Astr. Soc.*, **24**, pp. 325–90.
Sleep N. H. and **Sneel N. S.**, 1976. 'Thermal contraction and flexure of mid continent and Atlantic marginal basins', *Geophys. J. Roy. Astr. Soc.*, **45**, pp. 125–54.
Sugimura A. and **Uyeda S.**, 1973. 'Island arcs: Japan and its environs', *Development in Geotectonics*, **3**. Elsevier Publishing Company, Amsterdam, 247 pp.
Tapponier P., 1977. 'Évolution tectonique du système alpin en Méditerranée: poinçonnement et écrasement rigide-plastique', *Bull. Soc. Géol. Fr.* (7), **19** (3), pp. 437–60.
Treuil M. and **Varet J.**, 1973. 'Critères volcanologiques, pétrologiques et géochimiques de la genèse et de la differenciation des magmas basaltiques: exemple de l'Afar', *Bull. Soc. Géol. Fr.*(7), **15** (5–6), pp. 506–40.
Turcotte D. L. and **Oxburgh E. R.**, 1972. 'Mantle convection and the new global tectonics', *Ann. Rev. Fluid-Mech.*, **4**, pp. 33–68.
Vine F. J., 1968 'Magnetic anomalies associated with mid-ocean ridges', in R. A. Phinney (ed.), *The History of the Earth's Crust*. Princeton Univ. Press, pp. 73–89.
Walcott R. I., 1972. 'Gravity, flexure and the growth of sedimentary basins at a continental edge', *Geol. Soc. Am. Bull.*, **83**, pp. 1845–48.
Wyllie P. J., 1971. *The Dynamic Earth* (Textbook in Geoscience). Wiley and Sons, New York, 416 pp.

Index

This index is restricted to the themes and subjects dealt with in this book. The numbers in italics refer to the most important passages and figures. Figures are indexed, after the text references, by figure number and the page numbers (in parentheses) immediately following.